Astrophysical Jets
Open Problems

Edited by

Silvano Massaglia
Università degli Studi di Torino, Italy

and

Gianluigi Bodo
Osservatorio Astronomico di Torino, Italy

Gordon and Breach Science Publishers

Australia • Canada • China • France • Germany • India • Japan • Luxembourg •
Malaysia • The Netherlands • Russia • Singapore • Switzerland • Thailand

Amsteldijk 166
1st Floor
1079 LH Amsterdam
The Netherlands

British Library Cataloguing in Publication Data

Astrophysical jets : open problems
   1.Astrophysical jets
   I.Massaglia, Silvano II.Bodo, Gianluigi
   523'.01

ISBN 90-5699-637-1

Contents

CONTENTS   vii

Editors' Preface

This book is the result of a three-day meeting held in Torino, Italy, December 11–13, 1996. The main topics of this meeting involved stellar and galactic jets: their origin from young stars, binary systems and active galactic nuclei; their propagation in the different environments with the related effects on stability, matter entrainment, and turbulence generation; and finally, their termination to form Herbig-Haro objects, hot-spots and radio lobes. The general idea that we tried to pursue in organizing the meeting was to gather theoreticians and observers belonging to different communities (radio, AGN, stellar, and so on) but sharing a profound interest in jets, including objects with spatial scales spanning from hundreds of A.U. up to hundreds of kpc, and to encourage participants to present and discuss their still-in-progress research, with all the associated difficulties, i.e., *open problems.*

Extragalactic and stellar jets have revealed many of their mysteries since the coming of the latest generation of instruments, both space-based and on the ground. We know that, almost certainly, extragalactic jets move at relativistic speed, at least on parsec scale; we also know that the complex phenomenology of active galactic nuclei (AGN) can be unified in a comprehensive general scheme in which the jet's orientation with the line of sight is the basic parameter. We know that stars in their youth form jets that have velocities of several hundred km s^{-1} and emit shock-excited lines as they propagate into the parent molecular clouds. The jet phenomenon, therefore, is evident in young objects: young stars, nuclei of galaxies in their initial stages, high red-shift QSOs, and so on. But evolved objects also generate jets when in binary systems; think of SS433 or the recently discovered galactic superluminal sources, GRS 1915+105 and GRO J1655-40. There is a general consensus that, among this variety of astrophysical sources, the common requirement for generating jets is strong accretion through disks or tori.

However, many basic aspects of jets remain to be clarified:

1. their *composition*, whether ordinary matter or electron-positron pairs, with the obvious consequences on the energy budget of the sources;

2. their *acceleration mechanism*, in which magnetic field plays, quite likely, a fundamental role (but radiative effects also have to be explored, especially as far as relativistic jets are concerned);

3. their *confinement*, whether due to thermal pressure balance or to magnetic fields;

4. the *origin* of radiating particles in extragalactic jets, i.e., how they are *re-accelerated* in order to keep the spectral index nearly constant along the jet;

5. the difference between *FRI* and *FRII* radio sources, whether intrinsic to the central source or related to the different environments in which the jets propagate;

6. the *velocity* itself of extragalactic jets on the kpc scales and the relative Mach number, and how the jet slows down to non-relativistic velocities in FRI sources;

7. the *connection* between bipolar flows and the embedded YSO jets.

These problems were discussed intensively during the meeting, and the outcome of these discussions is presented in the 22 chapters of this volume. The contributions have been divided into four sections. The first section is devoted to the origin, acceleration and collimation of jets; we see that MHD effects are dominant even though one must pay considerable attention to radiative Compton drag for relativistic jets. The second section concerns AGN jets in the vicinity of their parent source, their composition and superluminal motions. The third section deals with the large scale of AGN jets, their confinement, the asymmetry problem and stability. The final section treats YSO jets, their connection with the surrounding bipolar flows, their emission properties and morphology.

In organizing the meeting we received financial support from the Osservatorio Astronomico di Torino, and we were hosted by the Rector of the University of Torino. We are grateful to Paola Rossi and Edoardo Trussoni for their continuous support, scientific and otherwise, before and during the meeting.

ORIGIN, ACCELERATION AND COLLIMATION

CHAPTER 1

ORIGIN, ACCELERATION AND FLARING OF JETS

M. CAMENZIND

Landessternwarte Königstuhl, D-69117 Heidelberg, Germany

It is widely accepted that the formation of jets in active galactic nuclei and in young stellar sources is ultimately related to the existence of gaseous discs around some central object. Rotating black holes are still thought to be the prime-mover behind the activity detected in centers of galaxies, while, in the case of protostellar jets, rapidly rotating stars and discs are responsible for the ejection of bipolar outflows. In both cases, magnetic fields are invoked for the acceleration and the collimation of these outflows. The basic elements of MHD driven relativistic disc winds are reviewed, and the question of acceleration and collimation is discussed. From these considerations it follows that the jets of young stellar objects and of black hole driven outflows can only be produced magnetically and that their propagation is determined by their magnetic properties. Such jets have low Mach numbers $\simeq 2 - 3$, and their instabilities are dominated by the pinch and kink modes. The structure in the collimated region is worked out, and it is shown that the relativistic jets are formed by a two-component plasma consisting of thermal ions and relativistic electrons. No thermal electrons are left over beyond passing the light cylinder. This plasma is still rotating leading to interesting emission properties. Knots closest to the core are attributed to compression by the time-dependent pinches and kinks.

1. INTRODUCTION

It is now known for long time that the cores of elliptical galaxies and young low-mass stellar objects are capable of driving highly colli-mated bipolar outflows. The velocities in the case of protostellar jets are in the range of a few hundred km per second, while the plasma flows driven by the cores of elliptical galaxies are relativistic having bulk Lorentz factors of at least a few, in some cases probably 100. Relativistic outflows are thought to be driven by rapidly rotating supermassive black holes sitting in the very center of these galaxies. Some of these masses have been pinned down by recent HST measurements, the mass in the central few parsecs of M 87 is observed to be $3 \times 10^9 \, M_\odot$.

In the following we discuss various scenarios which lead to the formation of collimated outflows. All these processes invoke axisymmetric magnetic fields rooted in some central rotating object. Rotation and magnetization are the driving elements for jet formation. These two elements generate a Poynting flux near the surface of these objects (discs or stars) which is then transformed into kinetic energy when the plasma moves along collimated flux surfaces. This theory is now completely understood, and can even be formulated on the background of relativistic objects such as rapidly rotating black holes and neutron stars. The only element that is still missing is a self–consistent treatment of outflow and collimation.

The collimation process is a consequence of magnetic plasma confinement. Axisymmetric self–pinched equilibrium solutions are well–known in plasma physics. A plasma configuration can be in equilibrium with its pressure forces, curvature forces and the pinch forces generated by currents flowing in the plasma. In astrophysics, interesting self–pinched plasmas occur in magnetically driven outflows generated by some rapidly rotating magnetized object. These configurations are thought to give rise to self–collimated outflows visible as jets in radio galaxies, quasars and protostellar objects. The magnetic field structures are generated by the underlying rapidly rotating objects (stars and accretion discs).

We discuss self–collimated jets as pinch solutions of the transverse force–balance equation. This results as a generalization of the classical pinch equilibrium, except that additional forces are generated by the rotating magnetic surfaces and the plasma streaming along these surfaces. Currents are self–induced and cannot be prescribed in an ad hoc fashion. Of particular importance is the centrifugal force exerted by the specific angular momentum of the streaming plasma. Without this force, the far–field would behave as a monopole. Inclusion of this force leads to a modified monopole that decays logarithmically weaker than a monopole. This has far–reaching consequences for the acceleration of the plasma along such flux tubes.

We also address the question of the nature of the plasma in the relativistic jets of quasars. This plasma probably consists of a normal plasma, including magnetic fields, thermal ions and relativistic electrons. This is a new plasma state which cannot be generated in the lab. It explains the absence of line emission in AGN jets as well as the reduction of Faraday depolarization required by these objects.

2. SCENARIOS OF JET FORMATION

Observationally, there are various constraints on the typical parameters for jet formation. HST data (Ray et al., 1996; Stapelfeldt et al., 1995) support models in which the outflow is initially poorly focused and then collimated into a cylindrical structure with diameters of at least of order of a few tens of AU. This is the most important information concerning the overall structure of young stellar jet formation. These scales correspond typically to a few thousand stellar radii. Protostellar jets are not pencil–like outflows from the inner edge of accretion discs, these are wide outflows, probably initially of spherical shape, until the flows are suddenly collimated into a cylindrical flow structure (Fig. 1).

Similar constraints follow from VLBI observations on extragalactic jets (Table 1). These parsec–scale jets appear already collimated within a fraction of a light year. This scale corresponds to at least a few hundred Schwarzschild radii, since the central masses cannot exceed $10^{10}\,M_{\odot}$. These geometric constraints are therefore the most severe restrictions on jet formation models. This also means severe implications on the dynamic range of numerical jet formation. In the framework of magnetohydrodynamic (MHD) models this implies solar wind type outflows from the inner part of the star–disc system with a collimation mechanism operating preferentially on the scale of a few thousand stellar radii in young stellar objects, and similarly for outflows driven by black holes. Also here, the outflow is first radial and is then collimated on a typical scale of a few hundred Schwarzschild radii into a cylindrical shape – despite many erroneous cartoons in the literature. Since these magnetospheres are rotating very rapidly – the corresponding light cylinder is only at a few injection radii, these outflows must be treated within a relativistic formulation.

Rapidly rotating objects have been proposed to be the ultimate sources of collimated outflows, once they carry a sufficiently strong magnetosphere. Since Keplerian discs represent the most extreme rotational states, they are *prime candidates* for jet drivers (Blandford & Payne 1982, and many papers thereafter). The crucial question here is whether the disc is able to maintain a *dipolar magnetic structure*. Quadrupolar field structures are unfavorable for driving efficiently disc outflows.

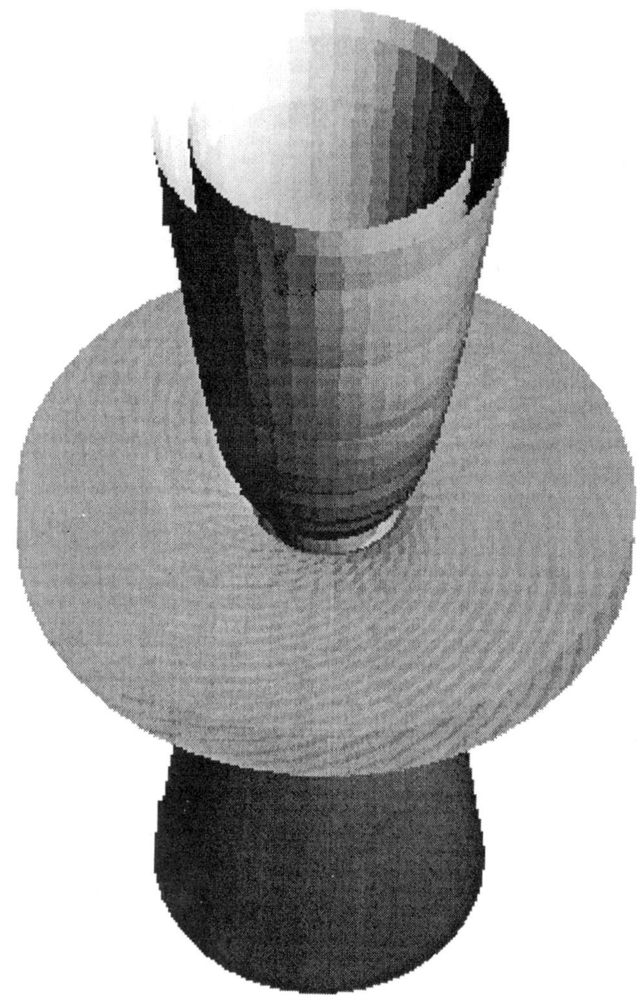

FIGURE 1. The essential elements of MHD jet formation: a central object, accretion disc and collimated rotating magnetic surfaces which guide the plasma flow. These magnetic surfaces are formed by magnetic field lines rooted either in the disc or the central object. These surfaces are collimated into a cylindrical shape at a certain radius, which is 10 – 100 AU for protostellar outflows and about one hundred Schwarzschild radii for black hole driven outflows in Quasars.

2.1. Jets from AGN Discs

We consider the interaction of an inward flow of material inside a disc with velocity $\vec{v}$ and the magnetic field in the disc. The discs near protostars are probably heavily convective and therefore prone to *dynamo action*. Advection and convective motion lead to a source for poloidal magnetic fields $\vec{B}_p = (\nabla\Psi \times \vec{e}_\phi)/R$ given in terms of the flux function $\Psi = RA_\phi$ (or the vector potential A_ϕ)

$$\frac{\partial\Psi}{\partial t} + \vec{v}\cdot\nabla\Psi - \eta R^2\,\nabla\cdot\left[\frac{1}{R^2}\nabla\Psi\right] = \alpha_d\,RB_\phi\,. \tag{1}$$

Poloidal magnetic flux Ψ undergoes a time–evolution due to the advection of the flux by means of the accretion process (with velocity $\vec{v}$) and diffusion due to turbulent processes, given by the magnetic diffusivity $\eta \simeq l_c v_c/3$ with convective velocity v_c on scales $l_c < H$. Regeneration out of toroidal magnetic fields B_ϕ, or its current function cRB_ϕ, occurs due to the dynamo effect represented by the function $\alpha_d \simeq v_c$. This toroidal field itself undergoes a similar time–evolution, written here for the current function $I = cRB_\phi$,

$$\frac{\partial I}{\partial t} + R^2\nabla\cdot\left[\frac{\vec{v}}{R^2}I\right] - R^2\,\nabla\cdot\left[\frac{\eta}{R^2}\nabla I\right] = R^2\vec{B}_p\cdot\nabla\Omega\,. \tag{2}$$

Differential rotation in discs mainly drives the evolution of the toroidal component, winding it up to equipartition field strengths. The magnetic diffusivity η is in general R–dependent and can therefore not be taken out of the diffusive term. When dynamo action is absent, $\alpha = 0$, the poloidal field structure evolves independently of the toroidal field and is only determined by initial and outer boundary conditions. The toroidal field is essentially determined by the radial component of the poloidal magnetic field. If there is no bending of field lines in the disc, toroidal fields cannot be generated. These two equations for the transport of magnetic flux and poloidal current flux in accretion discs can also be formulated on the background of rotating black holes (Khanna and Camenzind 1996).

The above equations describing the evolution of magnetic fields inside the disc, must be supplemented by equations for the behavior of the magnetic field outside the disc. Usually, vacuum conditions are assumed in the exterior region, i.e. no currents are flowing in the exterior region. If the magnetic fields are, however, loaded with disc plasma, the poloidal field structure has to be taken as a solution of the Grad–Shafranov equation (see later on), and currents are allowed to circulate between the disc surface and the wind region. This coupled

2D problem is quite complicated and has not yet been solved in the literature (for 1D solutions, see Ferreira, 1997).

For rapidly rotating discs, this dynamical system evolves naturally into a *quadrupolar structure* (Camenzind, 1990; Khanna and Camenzind 1996; Rüdiger et al., 1995), quite in analogy to the time evolution of galactic magnetic fields. It is quite important for the stability of the magnetic fields that accretion discs are highly diffusive. Simulations based on ideal MHD equations are not suitable to test the time evolution of magnetic fields in discs. Outflows driven by such configurations could stabilize these structures. But at the moment there is no way to couple numerically a diffusive disc with an ideal MHD flow exterior to the disc. Discs have extremely low Reynolds numbers, outflows high Reynolds numbers.

TABLE 1: A comparison of galactic and extragalactic jet sources

Parameters	Protostellar	Parsec–scale Quasars
Mass flow	$10^{-8\pm2}\ M_\odot\ yr^{-1}$	$0.0001 - 0.1\ M_\odot\ yr^{-1}$
Radii	$10 - 100$ AU	$0.001 - 1$ lyr
Density	$10^{3\pm2}\ cm^{-3}$	$10^{2\pm2}\ cm^{-3}$
Magnetic field	$\simeq$ milliGauss	$\simeq$ Gauss

2.2. Jets from Rapidly Rotating Axisymmetric Stars

Fields from discs can be flared up to larger radii, provided there is some extra pressure along the axis. The most natural source of such a pressure is an additional magnetosphere built up by the central star itself. Such configurations have been discussed by the present author for the extragalactic case already in 1986 (Camenzind 1986a) and applied to protostellar outflows in 1990 (Camenzind 1990). A somewhat similar model has been discussed by Shu et al. in 1994. Personally, I think that only the existence of such a central pressure can provide the observed large collimation radii in protostellar sources.

The strong differential rotation of accretion discs is responsible for the excitation of quadrupolar modes in discs. Since differential rotation is probably suppressed in protostellar objects, the original dipolar structure left over from the collapse process, could be stabilized over long time scales. This is still an open question, but this mechanism would provide a different scenario for jet formation (Camenzind 1990). Initially, we thought that the interaction between the stellar magnetosphere and the surrounding disc could drive the outflows (see also Fendt et al. 1995).

Such outflows could, however, also be driven by the *central object* itself. This idea has a long history and requires that the central object carries an axisymmetric magnetosphere. Jet outflows occur in magnetized astrophysical systems whenever they provide a perfect *axisymmetry*. Real three–dimensional objects such as pulsars and Weak–Line T Tauri stars never show the existence of collimated plasma flows.

3. THE QUESTION OF ACCELERATION

Though the solar wind is magnetized, it is not magnetically driven. Winds ejected from rapidly rotating stars and discs are, however, magnetically driven. When a solar type star reaches the main sequence with a rotation rate $\Omega_* \simeq 20\Omega_\odot$ at an age of 50 million years, then the thermally driven solar wind changes to a completely magnetically and rotationally driven MHD wind (MacGregor 1996). Historically, these winds have been discussed as spherically symmetric winds in the equatorial plane of a star that rotates uniformly with angular velocity Ω_*. In the following, we will discuss a fully 2D approach that can handle any form of the magnetic flux surface. This is important if one wants to model magnetized outflows from rapidly rotating objects.

3.1. What drives the Acceleration?

When a rotating body carries an axisymmetric magnetosphere, the rotation induces electric fields $\vec{E}_\perp$ over the unipolar induction such that the magnetic surfaces become equipotential surfaces. These electric fields together with the toroidal magnetic fields $\vec{B}_T$ give rise to a Poynting flux

$$\vec{P}_p = \frac{c}{4\pi}\,\vec{E}_\perp \times \vec{B}_T\,.\tag{3}$$

This can be written in the form, $I_* = -(cRB_\phi/2)_*$

$$\vec{P}_p = \frac{\Omega_* I_*}{2\pi c}\,\vec{B}_p\,.\tag{4}$$

When integrating this Poynting flux over the entire surface of the central object, we end up with a magnetic luminosity

$$L_{\mathrm{mag}} = \frac{1}{c}\,\Omega_* I_* \Psi_*\,,\tag{5}$$

depending on the rotational state of the object, the total magnetic flux Ψ_* carried by the magnetosphere and the total current I_* driven by the rotation of the object. This magnetic luminosity is a new source of energy which is usually neglected in the energy budgets. It is however an important source of energy for strongly magnetized and rapidly rotating objects.

In the case of supermassive black holes sitting in the center of bright elliptical galaxies, the field rotation is related to the angular velocity Ω_H of the horizon which is a function of the angular momentum parameter $a_H < 1$ and the radius r_H of the horizon

$$\Omega_* \leq \Omega_H = \frac{1}{2}\frac{a_H c}{r_H} \simeq 10^{-4}\,\mathrm{rad\,s}^{-1}\,\frac{10^9\,M_\odot}{M_H}. \tag{6}$$

This rotation rate together with the typical values for the magnetic flux Ψ_H covering a black hole and the typical current driven through the magnetosphere provides the following estimate for the magnetic luminosity of quasars

$$L_{\mathrm{mag}} \simeq 3 \times 10^{46}\,\mathrm{erg\,s}^{-1}\,\frac{10^9\,M_\odot}{M_H}\,\frac{\Psi_H}{10^{33}\,\mathrm{Gauss\,cm}^2}\,\frac{I_H}{10^{18}\,\mathrm{Amps}}. \tag{7}$$

This has to be compared with the kinetic luminosity of bright jets, such as the jet of 3C 273

$$L_j = (\gamma_j - 1)\,\dot{M}_j c^2 \simeq 5 \times 10^{46}\,\mathrm{erg\,s}^{-1}\,\frac{\dot{M}_j}{0.1\,M_\odot\,\mathrm{yr}^{-1}}\,\frac{\gamma_j}{10}. \tag{8}$$

The mass outflow is typically a few percents of the mass inflow, or the accretion rate that determines the overall bolometric luminosity of quasars. *Therefore, we see that the magnetic luminosity of accretion discs around supermassive black holes can easily explain the kinetic luminosity in jets.* How magnetic energy is transformed into kinetic energy will be discussed in the next section.

Similarly, young stellar objects at the age of a few hundred thousand years are observed to rotate faster than the Sun, $\Omega_* \simeq 20\,\Omega_\odot$ with $\Omega_\odot = 3 \times 10^{-6}$ rad s^{-1}, and to carry a stronger magnetosphere. Field strengths of the order of 1000 Gauss are consistent with observations. This corresponds to a magnetic luminosity of the rotating star

$$L_{\mathrm{mag}} = 0.1\,L_\odot\,\frac{\Psi_*}{10^{25}\,\mathrm{Gauss\,cm}^2}\,\frac{I_*}{10^{14}\,\mathrm{Amps}}\,\frac{\mathrm{days}}{P_*} \tag{9}$$

with a magnetic flux given in units of 10^{25} Gauss cm^2 and a current of 10^{14} Amperes. This is some fraction of the bolometric luminosity of low–mass stellar objects, but completely negligible in the case of the Sun due its low mean magnetic field and its low rotation rate. For young stellar objects, the magnetic luminosity is close to the kinetic luminosity found in the jets of these objects

$$L_{\text{jet}} = 0.1\, L_\odot \, \frac{\dot{M}_j}{10^{-8}\, M_\odot\, yr^{-1}} \left(\frac{v_j}{300\,\text{km}\,\text{s}^{-1}}\right)^2 . \qquad (10)$$

The magnetic luminosity of rapidly rotating protostellar sources can therefore afford for the jet luminosity found in young stellar outflows.

3.2. Plasma Motion along the Magnetic Flux-Tubes

Any vector $\vec{V}$ can be split into a poloidal component $\vec{V}_p$ and a toroidal component $\vec{V}_T$ in ϕ–direction. Since $\nabla \cdot \vec{B}_p = 0$, the poloidal magnetic field derives from a vector potential $\Psi(R, z)$

$$\vec{B}_p = \frac{1}{R}\left(\nabla\Psi \times \vec{e}_T\right). \qquad (11)$$

This function is a constant along poloidal field lines. Field lines of total magnetic field wind around the magnetic surfaces $\Psi = const.$

Stationary plasma flows are now given by a number of first integrals, called surface functions. In a quasi–Newtonian description gravity can be included by using the pseudo–Minkowskian metric

$$ds^2 = -\alpha^2\, dt^2 + h_{ik}\, dx^i\, dx^k , \qquad (12)$$

where h describes the 3–space metric of a Schwarzschild background and α is the red-shift factor between local and universal observers, including the Newtonian potential Φ, $\alpha \simeq 1 + \Phi$. The most general case of an axisymmetric rotating background metric is discussed in Camenzind (1986b; 1996). In the following we use units with $c = 1 = G$ and cylindrical coordinates (R, ϕ, z).

The flux–freezing condition implies the existence of a function $\eta(\Psi)$ such that

$$\alpha n\, \vec{U}_p = \eta(\Psi)\, \vec{B}_p . \qquad (13)$$

This describes the mass flux in a magnetic flux–tube (n is the particle density). Similarly, the poloidal part of the induction equation leads to the relation

$$U_\phi = \frac{\eta(\Psi)}{\alpha n}\, B_\phi + \frac{\gamma R \Omega_*}{\alpha} . \qquad (14)$$

γ is the Lorentz factor of the motion. The red-shift factor α accounts for the transformation between local and universal time t. This shows that plasma motion is the sum of a field aligned component and a rotation about the polar axis. The plasma moves along the magnetic surface, since the high conductivity forbids diffusion against the magnetic field in the stationary approximation.

Two additional quantities follow from axisymmetry and stationarity. Axisymmetry requires angular momentum conservation for the specific angular momentum U_ϕ of the plasma and the angular momentum stored in the magnetic field

$$\mu U_\phi - \frac{\alpha R B_\phi}{4\pi\eta} = L(\Psi)\,. \tag{15}$$

Stationarity implies the existence of a relativistic Bernoulli–integral

$$\mu\alpha\gamma - \frac{\alpha R \Omega_* B_\phi}{4\pi\eta} = E(\Psi)\,. \tag{16}$$

μ is the specific enthalpy of the plasma

$$\mu = mc^2 + \frac{\Gamma}{\Gamma - 1}\,K_0(\Psi)\,n^{\Gamma-1}\,, \tag{17}$$

for a polytropic equation of state. These conservation laws can easily be generalized to a Kerr background metric (Camenzind 1986b; 1996).

Due to these conservation laws, the Lorentz factor γ, the specific angular momentum U_ϕ of the plasma and the current RB_ϕ are given in terms of the Mach number M defined as $M^2 = \mu_0\mu\eta^2/n = \alpha^2 U_p^2/V_A^2$ with V_A as the Alfvén velocity (Camenzind 1986b, see also Eq. 51)

$$\alpha\gamma = \frac{E}{\mu}\,\frac{\alpha^2(1-\epsilon) - M^2}{\alpha^2 - M^2 - x^2} \tag{18}$$

$$U_\phi = \frac{E}{\mu}\,R_L c\,\frac{(1-\epsilon)x^2 - M^2\epsilon}{\alpha^2 - M^2 - x^2} \tag{19}$$

$$\alpha R B_\phi = \frac{4\pi\eta L}{\epsilon}\,\frac{x^2 - \alpha^2\epsilon}{\alpha^2 - M^2 - x^2}\,. \tag{20}$$

$x = R/R_L$ with $R^2 = g_{\phi\phi}$ is a special relativistic correction related to the light cylinder radius $R_L = c/\Omega_*$, and $\epsilon = \Omega_* L/E \leq 1$ measures the influence of the magnetic energy. The Alfvén point is now at the position where $M_A^2 = \alpha^2 - x_A^2$. As in Newtonian MHD flows, the regularity at the Alfvén point determines the total angular momentum L lost by the wind, $\alpha_A^2\epsilon = x_A^2$. For Newtonian flows, the parameter ϵ is extremely small, in the solar wind $\epsilon_\odot \simeq 10^{-8}$ (the light cylinder for solar rotation is at 700 AU). This parameter is only two orders of magnitude higher for outflows driven by rapidly rotating low-mass stars. This parameter approaches, however, unity, whenever the flows become relativistic. For this reason, the corrections of order ϵ cannot be neglected for relativistic outflows.

The current I carried by MHD winds is essentially determined by the total angular momentum lost through the wind, $I \propto \eta L$. The expression for RB_ϕ can be written, by using $M^2 = 4\pi\mu\eta^2/n$ and $U_p = (\eta/\alpha n)B_p$ in the form

$$I = \alpha R B_\phi = \Omega_*(B_p R^2) \frac{M^2}{x^2\beta_j} \frac{x^2 - x_A^2}{1 - M^2 - x_A^2} = -\Omega_*(B_p R^2) \frac{1}{\beta_j} i(x),$$

(21)

or as

$$I = -\Omega_* (B_p R^2)_* \frac{1}{\beta_p} \frac{i(x)}{\Phi_\Psi(x)}$$

(22)

with the flux tube function Φ_Ψ as defined in Eq. (31). This relation demonstrates that the current in the jet is produced by the rotation of the central object and the magnetic flux contained in the flux tube.

Though these relations are now established for ten years, most investigations on hydromagnetic winds are still based on the Newtonian expressions. I am always surprised to see relativistic outflows resulting from purely Newtonian calculations!

This model of hydromagnetic flows includes special relativistic effects, as well as gravitational effects due to a central point source. It includes four surface functions η, L, E and K_0, the latter is a measure of the pressure at the foot point of the flux surface. When magnetospheres are differentially rotating, also the field line rotation Ω_* is a function of the flux surface. Not all of these constants are independent. Three of them follow from regularity conditions required at the critical points of the plasma flow.

3.3. The Wind Equation

The structure of the plasma flow is only known, when the Mach number is given along a prescribed flux surface. This equation is known as the *wind equation* (Camenzind 1986a,b; 1996; Camenzind and Englmaier 1997). In Newtonian MHD it follows from the Bernoulli integral (Heyvaerts and Norman 1996), in the relativistic version it is a consequence of the normalization of the 4–velocity, $U_\beta U^\beta = -1$, or

$$\alpha^2\gamma^2 - \vec{U}^2 = 1 \quad , \quad \vec{U} = \gamma\vec{v}$$

(23)

which leads to the equation

$$\vec{U}_p^2 + 1 = \alpha^2\gamma^2 - \frac{U_\phi^2}{R^2}.$$

(24)

Using the solutions for γ and U_ϕ, this can be arranged into the form

$$\vec{U}_p^2 + 1 = \left(\frac{E}{\mu}\right)^2 U_{\mathrm{M}}(R; M^2, \epsilon) \tag{25}$$

with

$$U_{\mathrm{M}}(R; M^2, \epsilon) = \frac{F_{\mathrm{M}}(R; M^2, \epsilon)}{x^2 \, D_{\mathrm{M}}^2} \tag{26}$$

and the definitions of two dimensionless functions

$$D_M \equiv \alpha^2 - M^2 - x^2 \tag{27}$$

$$F_M \equiv x^2[\alpha^2(1 - \epsilon) - M^2]^2 - [(1 - \epsilon)x^2 - M^2\epsilon]^2 \,. \tag{28}$$

With the relation between $\vec{U}_p$ and $\vec{B}_p$

$$U_p^2 = \frac{\eta^2 B_p^2}{\alpha^2 n^2} = \frac{B_p^2}{16\pi^2\mu^2\alpha^2\eta^2} \, M^4 \,, \tag{29}$$

the above equation represents a quartic relation in M for each position along a flux surface.

Multiplying the equation by $\alpha^2 x^4$ we obtain

$$\alpha^2 x^2 \, \frac{F_M'(x; M^2, \epsilon)}{D_M^2(x; M^2)} \left(\frac{E}{\mu}\right)^2 = \alpha^2 \, x^4 + \frac{B_p^2 x^4}{16\pi^2\mu^2\eta^2} \, M^4 \,. \tag{30}$$

The last term can be written in terms of the flux function

$$\Phi_\Psi^{-1}(x) \equiv \frac{B_p R^2}{B_{p*} R_*^2} \,, \tag{31}$$

which is normalized by the footpoint R_* and its magnetic field strength B_{p*}. In addition, we introduce the *dimensionless magnetization parameter*

$$\sigma_*(\Psi) \equiv \frac{(B_{p*} R_*^2)(\Psi)c}{4\pi\mu\eta(\Psi)R_L^2(\Psi)} \,. \tag{32}$$

This altogether leads to the equation

$$\alpha^2 x^2 \, \frac{F_M(x; M^2, \epsilon)}{D_M^2(x; M^2)} \left(\frac{E}{\mu}\right)^2 = \alpha^2 \, x^4 + \sigma_*^2 \, \Phi_\Psi^{-2} \, M^4 \,. \tag{33}$$

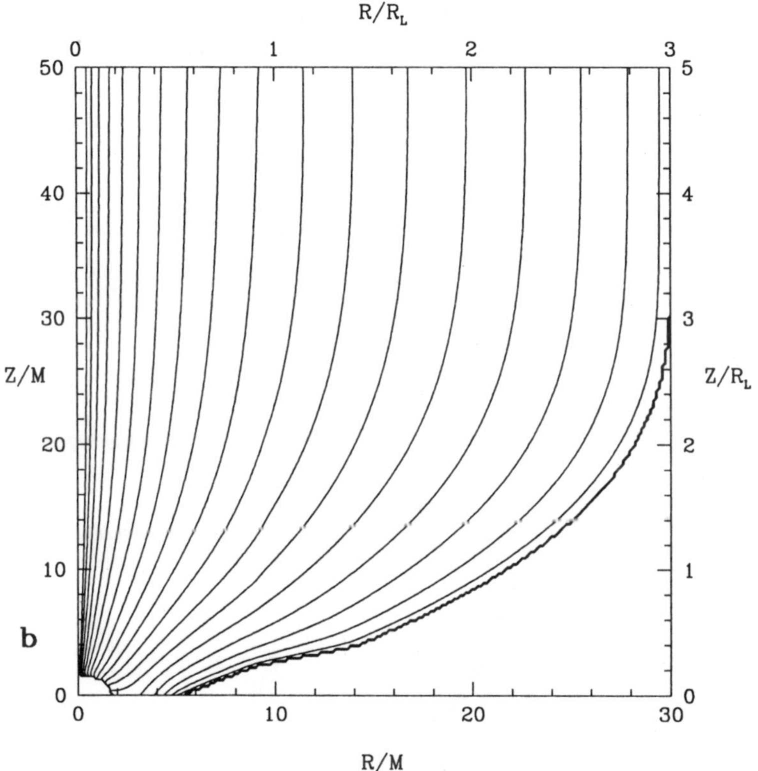

FIGURE 2. This figure shows the form of force–free flux tubes generated near a rapidly rotating black hole (Fendt 1997).

This fundamental equation clearly shows that the solutions of the wind equation, when formulated for the Mach number, depend on the following parameters

- the angular momentum parameter ϵ,
- the dimensionless energy $\bar{E} = E/mc^2$,
- the magnetization parameter σ_*,
- the flux tube function Φ_Ψ
- one additional parameter hidden in the specific enthalpy

$$\mu = mc^2 \left[1 + \frac{c_{S*}^2}{\Gamma - 1} \left(\frac{M_*^2}{M^2} \right)^{\Gamma - 1} \right]. \tag{34}$$

These are essentially 4 parameters for each flux surface $\Psi = const$ with known flux function Φ_Ψ (see Fig. 2 as an example).

Relativistic MHD Wind

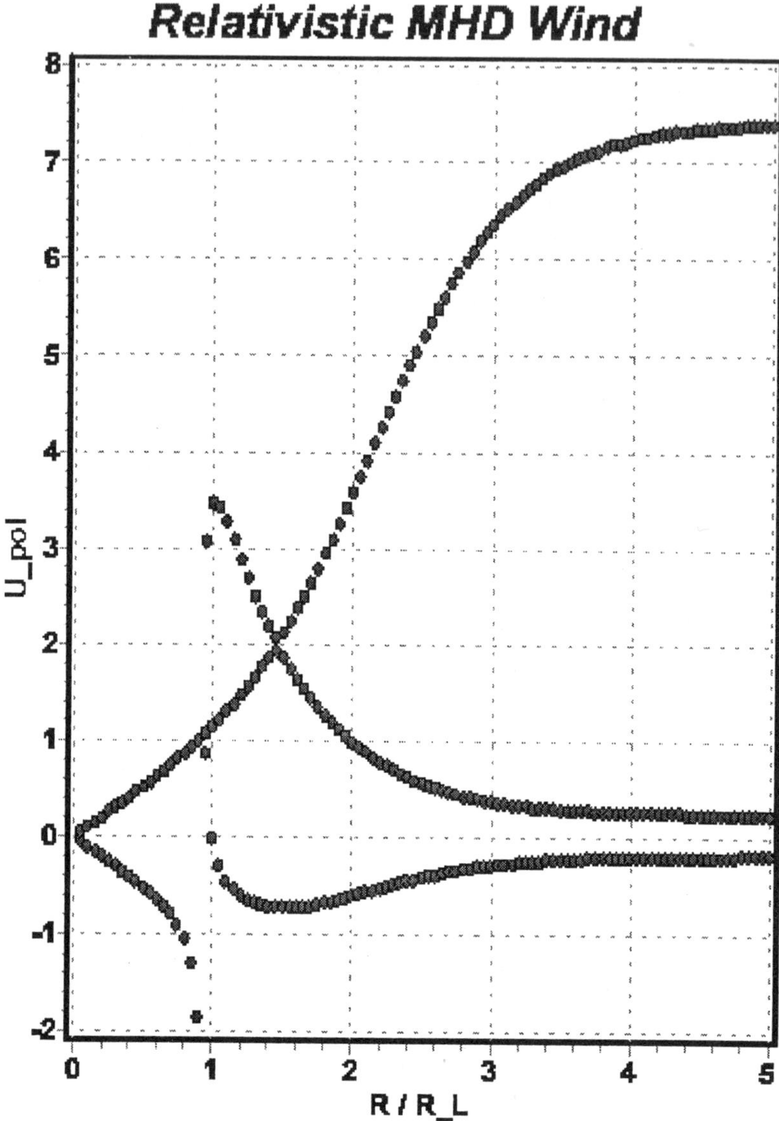

FIGURE 3. The various solution branches of the cold wind equation for a magnetization $\sigma_* = 10$ achieved in a flux-tube collimated into a cylindrical shape beyond the light cylinder (Fig. 2). The wind starts essentially with vanishing velocity, passes the Alfvén point at $R \simeq R_L$, crosses the fast magnetosonic point (X-point) and finally saturates to a constant plasma motion. The other solution branches are unphysical.

It can be shown that the relativistic wind equation has the same critical points as the non–relativistic one: the slow magnetosonic point, the Alfvén point $D_M(R_A) = 0$ and the fast magnetosonic point (Fig. 3). It is very important in this respect that the light cylinder is not a critical point of the wind equation.

The requirement that a wind solution passes through all three critical points fixes therefore three of the four parameters σ_*, L, E and M_*. We may consider σ_* as a free parameter which is fixed by the disc physics.

Solutions of the wind equation can easily be obtained by using the expression for the Mach number

$$M^2 = \frac{\mu}{mc^2}\, U_p x^2\, \frac{\Phi_\Psi}{\sigma_*}\,. \tag{35}$$

For normal plasma outflow (consisting of ions and electrons), the sound velocity is not important so that flows can be approximated by cold winds. This is equivalent to saying that the slow magnetosonic point is identical with the injection point. In this case, the wind equation reduces to a fourth–order polynomial in the poloidal velocity

$$A_4 U_p^4 + A_3 U_p^3 + A_2 U_p^2 + A_1 U_p + A_0 = 0 \tag{36}$$

with coefficients given in Camenzind (1986b) (see also Fendt and Camenzind 1996)

$$A_0 = x^2 (1 - x^2)^2 - E^2\, x^2 (1 - x^2)(1 - x_{\mathrm{A}}^2)^2\,, \tag{37}$$
$$A_1 = -2\, g\, x^4 (1 - x^2 - E^2 (1 - x_{\mathrm{A}}^2)^2)\,, \tag{38}$$
$$A_2 = g^2\, x^4 (x^2 - E^2 (x^2 - x_{\mathrm{A}}^4)) + x^2 (1 - x^2)^2\,, \tag{39}$$
$$A_3 = -2\, g\, x^4 (1 - x^2)\,, \tag{40}$$
$$A_4 = g^2\, x^6\,. \tag{41}$$

Here we use the quantity $g \equiv (\Phi_\Psi/\sigma_*)$.

This form is extremely useful, since it immediately produces all branches of solutions for a given flux tube function $\Phi_\Psi(x)$ and given magnetization σ_*. A particular solution of this cold wind equation (for $\alpha = 1$) is shown in Fig. 3 for a flux-tube function of the form represented by Fig. 2 (a force–free solution of the Grad–Shafranov equation). Here, the magnetization $\sigma_* = 10$, the initial Poynting flux is converted into kinetic energy essentially beyond the light cylinder. As shown in Fig. 4, about 80% of the Poynting flux is converted into kinetic energy, the asymptotic Lorentz factor is $\simeq 8$.

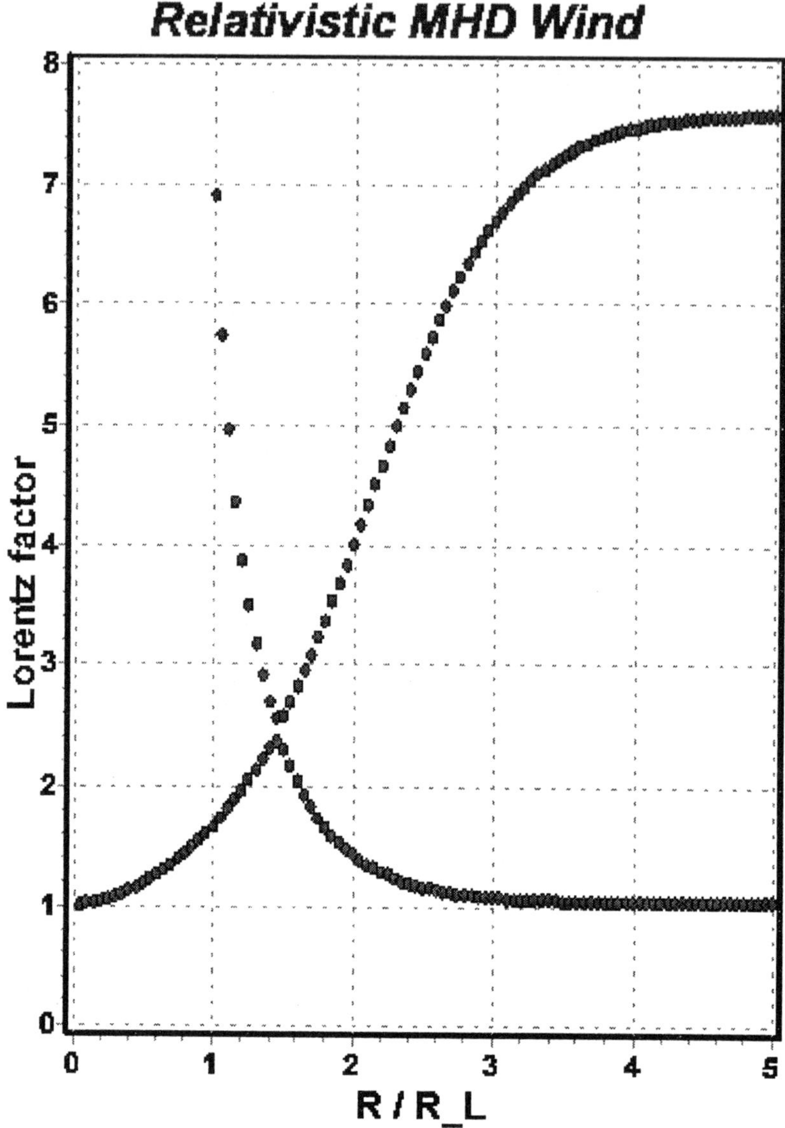

FIGURE 4. The behavior of the Lorentz factor along a given flux tube for the solution shown in Fig. 3. This demonstrates that asymptotic Lorentz factors of the order of 10 can be achieved for magnetizations of the same order of magnitude.

Of special interest is the evolution of the specific angular momentum in the plasma (Fig. 5). Similar to the conversion of magnetic energy into kinetic energy of the plasma, there is also a conversion of angular momentum carried the magnetic field into angular momentum of the plasma.

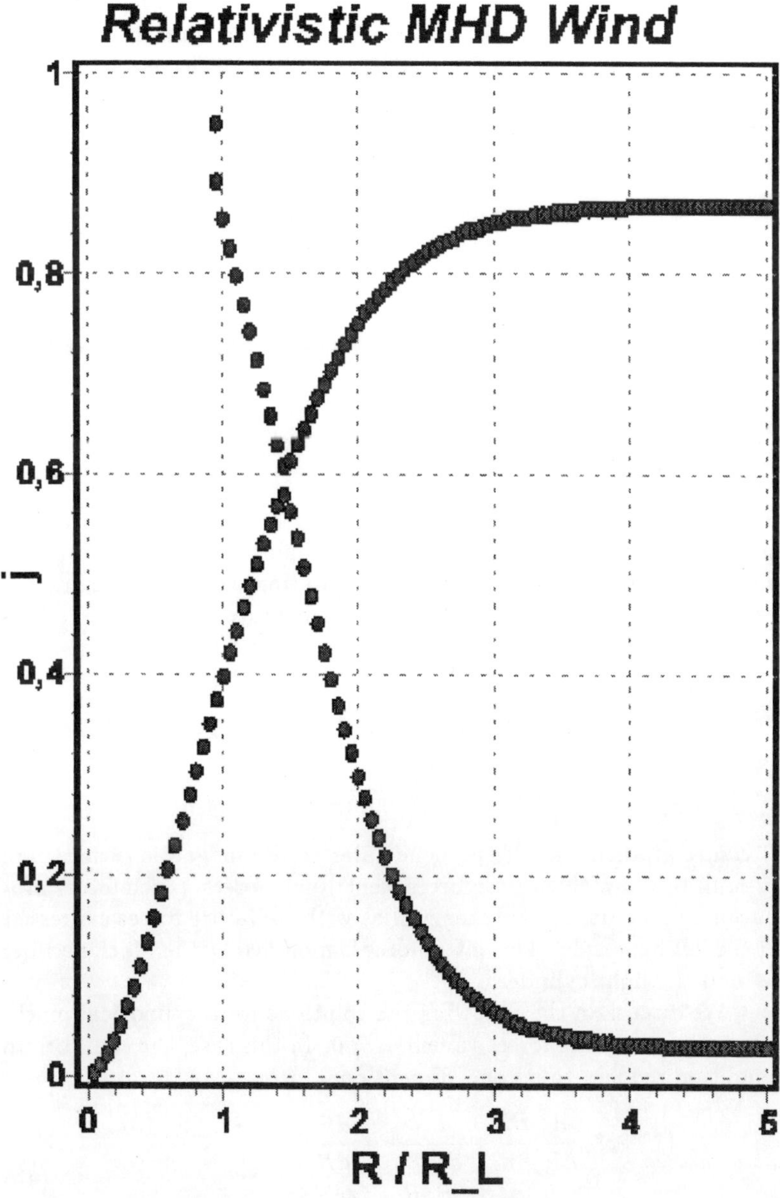

FIGURE 5. The behavior of the specific angular momentum j along a given flux tube for the solution shown in Fig. 3, in natural units of $R_L c$. For $\sigma_* \gg 1$, $j \to R_L c$.

As shown in Fig. 5, this conversion occurs very efficiently near the light cylinder. For high magnetization, $\sigma_* \gg 1$, the specific angular momentum tends towards the asymptotic limit R_{LC} given by the light cylinder. Even for a moderate magnetization, $\sigma_* = 10$, this limit is achieved within 80%. This is a very important result telling us that plasma outside the light cylinder is still rotating with a specific angular momentum roughly given by R_{LC} in Quasar jets.

4. MHD JETS AS PINCH SOLUTIONS

The magnetic structure of jets is quite universal. The entire jet consists of a family of nested magnetic surfaces (Fig. 1), and plasma is moving along these surfaces.

4.1. The Transverse Equilibrium

The collimation of magnetized winds into a conical outflow is very similar to the plasma confinement in a z–pinch (Meier et al. 1996). The equilibrium condition for a non–rotating axisymmetric current carrying pinch configuration has to be generalized to include various relativistic effects (Appl and Camenzind 1993a,b)

$$\kappa \frac{B_p^2}{4\pi}\left(1 - M^2 - x^2\right) = \left(1 - x^2\right)\nabla_\perp \frac{B_p^2}{8\pi} + \nabla_\perp \frac{B_\phi^2}{8\pi} + \nabla_\perp P$$
$$-\frac{B_p^2 \Omega^F}{4\pi c^2}\,\nabla_\perp(\Omega^F R^2) + \left(\frac{\mu n j^2}{R^3} - \frac{I^2}{4\pi R^3}\right)(-\nabla_\perp R)\,. \tag{42}$$

Pressure gradients acting perpendicular to the magnetic surfaces are in equilibrium with electric forces, centrifugal forces, pinch forces produced by poloidal currents, as well as with curvature forces expressed at the left hand side. This latter force is modified by the Mach number M and the light cylinder.

We discuss in the following the solutions for a cylindrical pinch, where the curvature forces vanish, $\kappa = 0$. In this case, the equilibrium condition reduces to its one–dimensional form

$$\left(1 - x^2\right)\frac{d}{dR}\frac{B_p^2}{8\pi} + \frac{d}{dR}\frac{B_\phi^2}{8\pi} + \frac{dP}{dR}$$
$$-\frac{B_p^2 R}{2\pi R_L^2} - \left(\frac{\mu n j^2}{R^3} - \frac{I^2}{4\pi R^3}\right) = 0\,. \tag{43}$$

The pressure of the toroidal field and the pinch force can be combined into one single expression. In addition, we normalize radii in units of the light cylinder R_L, $x = R/R_L$, and the poloidal magnetic field in units of its value on the central axis, B_0, $y = B_p^2/B_0^2$,

$$(1 - x^2)\frac{dy}{dx} - 4xy + \frac{1}{x^2}\frac{dI^2}{dx} + \frac{8\pi}{B_0^2}\frac{dP}{dx} - \frac{8\pi\mu n}{R_L^2 B_0^2}\frac{j^2}{x^3} = 0. \tag{44}$$

Electric fields due to the rotation of the magnetic surfaces modify the classical pinch equilibrium in two ways. First, the equation has a singular point at the light cylinder, and secondly, the additional term $4xy$ is crucial for the form of the equilibrium. In contrast to force–free models (Appl and Camenzind 1993b), the specific angular momentum j of the plasma and the poloidal current I are not arbitrary, but follow from disc winds, as discussed in the previous section.

4.2. Simple Pinch Solutions

We first discuss solutions of the relativistic pinch equilibrium neglecting the influence of the centrifugal force. In this case the equilibrium is determined by the condition

$$(1 - x^2)\frac{dy}{dx} - 4xy + \frac{1}{x^2}\frac{dI^2}{dx} + \frac{8\pi}{B_0^2}\frac{dP}{dx} = 0. \tag{45}$$

The toroidal current follows to a good approximation, $i = 1$,

$$I = -\Omega^F(R^2 B_p)\frac{1}{\beta_p} = -B_0 R_L\left(x^2\sqrt{y}\right)\frac{1}{\beta_p} \tag{46}$$

and we prescribe the plasma pressure in terms of $P = \beta B_p^2/8\pi$ with a constant plasma beta. This leads finally to the equation, $\gamma_p = 1/\sqrt{1 - \beta_p^2}$,

$$\left(1 + \beta + \frac{x^2}{\gamma_p^2\beta_p^2}\right)\frac{dy}{dx} + \frac{4}{\gamma_p^2\beta_p^2}xy = 0. \tag{47}$$

This homogeneous equation has the remarkably simple solution

$$y(x) = \frac{C}{(1 + \beta + x^2/x_c^2)^2} \tag{48}$$

with

$$x_c = \gamma_p\beta_p \quad , \quad R_c = \gamma_p\beta_p R_L \tag{49}$$

as the *core radius* of the jet. Inside the core, given by the light cylinder radius and the poloidal velocity U_p, the poloidal magnetic field is essentially homogeneous and decays as a monopole field outside the core, $B_p \propto 1/R^2$ for $R > R_c$

$$B_p = \frac{B_0}{1 + \beta + R^2/R_c^2}\,. \tag{50}$$

This is not unexpected since the jet is formed by a supermagnetosonic wind. The current function increases quadratically inside the core and saturates towards a constant value outside the core, $I_\infty \simeq cB_0 R_L \beta_p$. This is exactly the structure of the force–free solutions found previously. The constant C can be absorbed into the central value B_0.

The structure function $i(x)$ will modify this behavior somewhat. A stronger modification follows from the inclusion of the centrifugal force in the force–balance equation

$$\left(1 + \beta + \frac{x^2}{\gamma_p^2 \beta_p^2}\right) \frac{dy}{dx} + \frac{4}{\gamma_p^2 \beta_p^2}\, xy - \frac{2}{R_L^2 V_A^2}\, \frac{j^2}{x^3} = 0\,. \tag{51}$$

The Alfvén velocity V_A is defined as $V_A^2(x) = B_0^2/4\pi\mu n$. This is now an inhomogeneous equation which can be solved in terms of the ansatz

$$y(x) = \frac{\gamma(x)}{(1 + \beta + x^2/x_c^2)^2} \tag{52}$$

where $\gamma(x)$ satisfies the equation

$$\frac{d\gamma}{dx} = \frac{2c^2}{V_A^2(x)}\, \frac{\bar{j}^2}{x^3}\, (1 + \beta + x^2/x_c^2) \tag{53}$$

with the dimensionless specific angular momentum defined as $j = R_L c\,\bar{j}$. The general solution will not be discussed in this context. This has an important consequence on the question of relativistic outflows. In a pure monopole field, the flux tubes converge too fast and lead to stalled plasma outflows for high magnetization (Fig. 2; Fendt and Camenzind 1996). Jet models including the centrifugal force are wider than the force–free solutions – the centrifugal force opens up the jet width. In addition, the current RB_ϕ is no longer leveling off at a constant value, but is increasing logarithmically, $I \propto \sqrt{\ln(R/R_c)^\alpha}$, beyond the core radius. The particular behavior depends of course on the functions $\beta_p(R)$, $M^2(R)$ and $i(R)$ which are only known if the global wind solution has been constructed. The above discussion just describes the gross features which are expected to occur in such a global solution.

Relativistic jets always have a core–envelope structure, along the axis the magnetic field is predominantly longitudinal, beyond the core *it becomes dominated by the toroidal component. The light cylinder $R_L \simeq 10 GM_H/c^2$ is the basic scale for Quasar jets, $R_L \simeq 30 - 100$ AU the basic scale for protostellar jets.*

4.3. On the Nature of Plasma in Relativistic Jets

Until now we have worked with a one–component description of the jet plasma, ions and electrons are tightly coupled and obey a Maxwellian distribution. The plasma in AGN jets is, however, extremely thin so that the tight coupling between electrons and ions is not guaranteed. We probably need a two–component description for the jet plasma. The ions will remain mostly thermal, but the electrons are probably boosted to a relativistic distribution near the light cylinder (Gangadhara and Lesch 1997). This would be an interesting effect – electrons will no longer be thermal in the collimated regime of the jet flow. This is consistent with observations which show that there is probably no thermal plasma present in the cores of jet sources. Electrons are easily accelerated to Lorentz factors of a few thousand. These electrons can be boosted up to even higher energies in reconnection zones, and their maximal energy is then essentially limited by radiative losses.

5. ROTATION OF THE JET PLASMA AND PLASMA DIAGNOSTICS

Synchrotron emission and inverse Compton emission from the parsec–scale jets can be used as a tool of plasma diagnostics for these jets. There are various interesting aspects of this emission which are shortly discussed.

5.1. Rotation of the Jet Plasma

With the solution of the specific angular momentum U_ϕ in the jet, we also get the angular velocity of the jet plasma, $\alpha = 1$,

$$R\Omega = \frac{U_\phi}{\gamma} = R_L c \frac{(1 - \epsilon)x^2 - M^2\epsilon}{1 - \epsilon - M^2} = R_L c \bar{j}. \tag{54}$$

From this we can derive the rotation period

$$P(R) = \frac{2\pi R^2}{R_L c} \bar{j}^{-1}(x). \tag{55}$$

Using the expression for the rotation of the black hole, Eq. (6), this can be written as, $\Omega_* = \omega \Omega_H$,

$$P(R) = \omega \, a_H \pi \, \frac{M_H}{c} \left(\frac{R}{M_H} \right)^2 , \qquad (56)$$

or as

$$P(R) \simeq 10 \, \mathrm{yr} \, \frac{\omega a_H M_H}{2 \times 10^9 \, M_\odot} \left(\frac{c^2 R}{100 \, G M_H} \right)^2 . \qquad (57)$$

As expected, the rotation period of the plasma in the jet scales with the mass of the black hole. Due to angular momentum conservation beyond the Alfvén point, it also scales with the square of its distance from the axis. For black holes in giant ellipticals, the plasma rotation is typically in the range of a few hundred years, while in lower masses it is of the order of a few years (BL Lac objects).

This period is, however, not the observed period. Due to light travel time effects, the period appears shortened for the observer (Camenzind and Krockenberger 1992)

$$P_{\mathrm{obs}} = (1 + z)\,(1 - \beta_j \cos\theta)\, P(R) \simeq \frac{(1+z)}{2} \, \frac{P(R)}{\gamma_j^2} \qquad (58)$$

for small inclination angles θ between the jet axis and the line of sight. For $\gamma_j \simeq 10$, this gives a period of the order of a few months in bright ellipticals, and a few days for moderate ellipticals with black hole masses $M_H \simeq 10^8 \, M_\odot$. When bright spots, or knots are propagating along collimated jets, the underlying rotation would produce a modulation of the light curve with the above period. Since the jets are slightly opening, the periods increase with time. We think that this kind of phenomenon is behind the regular structure of the optical light curve of some quasars and BL Lac objects (Schramm et al. 1994; Wagner et al. 1995). This modulation is not necessarily visible in the radio light curve, since optical depth effects smear out the modulations. The transition can be nicely seen in the mm–light curves observed for 3C 273.

This is one of the most fascinating aspects of collimated MHD jets on the parsec–scale of elliptical galaxies. This effect really depends on the magnetic effects of disc winds. The effect cannot be observed in the slow jets of protostellar objects. The rotation is too slow at distances of a few hundred Alfvén radii. It is probably indirectly visible in the line emission that is generated near the central object.

5.2. Polarization of the Synchrotron Emission

Spindeldreher (1996) has calculated the polarization properties of the synchrotron emission for the above jet models. The polarization vector is defined by

$$\vec{p} = \hat{n} \times \vec{B} \,, \tag{59}$$

since the radiation is polarized perpendicularly to the line of sight and the projected magnetic field. The degree of polarization is given by

$$m = \frac{\sqrt{q^2 + u^2}}{i} = \frac{i_p}{i} \,. \tag{60}$$

i and i_p are the total, respectively polarized specific intensities. In a rotating jet, however, an electric field exists in the observer frame Σ', while in the local plasma rest frame Σ the electric field vanishes. The normalized polarization vector $\vec{p}\,'$ follows then from a Lorentz transformation of the electromagnetic radiation field

$$\vec{p}\,' = \frac{\vec{E}'_\perp + \hat{n}' \times \vec{B}'}{\left| \vec{E}'_\perp + \hat{n}' \times \vec{B}' \right|} \,, \tag{61}$$

where

$$\vec{E}'_\perp = \vec{E}' - (\hat{n}' \vec{E}') \, \hat{n}' \tag{62}$$

is the projected electric field of the jet in the frame of the observer. One interesting property is that the radiation is in general not perpendicularly polarized to the magnetic field, as measured by an observer, contrary to a non-moving source. The polarization vector is a–priori not a measurement for the direction of the magnetic field due to aberration effects.

Polarization properties of the synchrotron radiation in the case of a cylindrically symmetric jet have been worked out under the assumption that in the plasma rest frame the density of electrons is constant. The polarization angle for narrow jets ($R_{jet} = 2R_c$) shows two possibilities:

- For small inclination angles of the jet the radiation will be polarized perpendicularly to the jet axis.
- For larger inclination angles, the radiation is polarized parallel to the beam axis. At the critical inclination, where the polarization angle flips from perpendicular to parallel polarization, the polarized intensity vanishes ($|m| = 0$).

This result is straightforward: when we are looking exactly along the axis, all polarization is washed out. For increasing inclination angles, also the polarized intensity increases until the line of sight is parallel to the velocity vector (the maximum in intensity). From here on, polarization decreases and vanishes again at a critical angle which is about twice the angle of maximum intensity. Beyond this critical angle, the polarization is always observed to be parallel to the jet axis, since aberration effects are now negligible. This is possibly the reason for the observational fact that the magnetic field of quasar jets is observed to be parallel to the jet axis. Since superluminal motion is maximal, we observe these jets under inclination angles smaller than twice the maximal one, polarization is perpendicular to the jet axis. The jets of BL Lac objects have Lorentz factors of $\simeq 4$ and are predominantly seen under high inclination angles. Their polarization is therefore observed to be parallel to the jet axis ('magnetic field perpendicular').

This behavior is also a function of the jet radius. In the case of large jet radii ($20R_L$), the radiation is only polarized parallel to the jet axis. The critical angle decreases with increasing jet velocity u_∞.

5.3. On γ-ray emission

Many Quasars and BL Lac objects are sources of high energetic gamma emission. This is due to inverse Compton cooling of the relativistic electrons in the parsec–scale jets (Dreissigacker 1996). Time-dependence is generated by means of the lighthouse effect and flaring activity due to reconnection e.g. Simultaneous observations from the near infrared into the TeV–region are very important to pin down the relevant physical processes.

6. FLARING OF JETS

Besides the steady plasma ejection from the disc corona, B_ϕ can wind up in the disc to virial values on a few diffusion time–scales, producing a magnetic pressure explosion and pushing material upward (called CDMEs, *coronal disc mass ejections*). As shown by the solar corona, coronal mass ejections are driven by sheared magnetic configurations. Based on this analogy, we propose a model for the synchrotron and gamma–ray flares observed in quasars and BL Lac objects. The stationary dipolar magnetic configuration is superseded by a current sheet formed inside expanding loops. Reconnection in the current sheet accelerates electrons to high energies. A hot plasmoid is formed inside the current sheet which is ejected and escapes along the jet at relativistic speeds. Since the plasmoid has a non–axisymmetric structure, the rotation along the underlying jet produces a complicated light curve (Dreissigacker 1996; Schramm et al. 1993; Wagner et al. 1995), which is probably a superposition from contributions by many small knots arranged in a banana–like configuration. On the mas–scale, this helical motion of synchrotron emitting knots in a background jet can be traced in many examples using mm–VLBI. This lighthouse model of blobs of relativistic electrons propagating with the speed of the jet explains the emission modulation by means of the helical trajectory of the blob within the jet. When the jet opens beyond the core radius of the galaxy, the rotation freezes out. While a few revolutions can be observed in the synchrotron emission, this is not necessarily true for inverse Compton cooling due to the rapid decay of the external photon density in the core of the galaxy.

7. CONCLUSIONS

The basic elements of formation, collimation and instabilities of relativistic MHD jets are essentially understood in a stationary approach to the problem. We can relate the outflow velocities to the inner boundary conditions, relativistic outflows naturally occur for strongly magnetized plasmas near a black hole. Collimation is also a consequence of the rapid rotation, since this rotation drives the currents which are necessary for the hoop stresses.

There is no doubt that protostellar, as well as Quasar outflows are collimated by magnetic processes. The main difference between hydro jets and MHD jets is the low Mach number of MHD jets. In these jets, the fast magnetosonic speed is the relevant signal speed which is close to the speed of light in Quasar jets. Even for protostellar jets, the magnetosonic speed is about 100 km/s and exceeds therefore by far the sonic speed. This is the ultimate reason why these jets have to be treated within MHD.

The basic element that is missing for the future is the time–dependent treatment of the jet formation problem. For this we need a jet–lab that can handle general relativistic MHD, relativistic hydro calculations are neither able to solve the jet formation problem, nor can they explain the observed properties of large–scale jets. The inclusion of magnetic and electric fields is essential for an understanding of quasar jet structures.

Acknowledgements

Part of this work is supported by the Deutsche Forschungsgemeinschaft (SFB 328).

REFERENCES

Appl, S., & Camenzind, M., 1993a, *A&A*, **270**, 71.

Appl, S., & Camenzind, M., 1993b, *A&A*, **274**, 699.

Blandford, R.D., & Payne, D.G., 1982, *MNRAS*, **199**, 883.

Camenzind, M., 1986a, *A&A*, **156**, 137.

Camenzind, M., 1986b, *A&A*, **162**, 32.

Camenzind, M., 1987, *A&A*, **184**, 341.

Camenzind, M., 1990, in *Reviews of Modern Astronomy* **3**, ed. G. Klare, Springer–Verlag (Heidelberg), p. 234.

Camenzind, M., 1993, in *The Jets of Radio Galaxies*, eds. H.J. Röser & K. Meisenheimer, Lecture Notes in Physics **421**, Springer-Verlag (Heidelberg), p. 109.

Camenzind, M., 1995, in *Reviews of Modern Astronomy* **8**, ed. G. Klare, Astron. Gesellsch. (Hamburg).

Camenzind, M., 1996, in *Solar and Astrophysical Magnetohydrodynamical Flows*, ed. K.C. Tsinganos, Kluwer, p. 699.

Camenzind, M., & Krockenberger, M., 1992, *A&A*, **255**, 59.

Camenzind, M., & Englmaier, P., 1997, *A&A*, submitted.

Dreissigacker, O., 1996, *Strahlungsmechanismen in Jets von Blazaren*, PhD thesis, University of Heidelberg.

Fendt, C., 1997, *A&A*, **319**, 1025.

Fendt, C., Camenzind, M., & Appl, S., 1995, *A&A*, **300**, 791.

Fendt, C., & Camenzind, M., 1996, *A&A*, **313**, 591.

Ferreira, J., 1997, *A&A*, in press.

Gangadhara, R.T., & Lesch H., 1997, *A&A*, in press.

Heyvaerts, J., & Norman, C.A., 1996, in *Solar and Astrophysical Magnetohydrodynamical Flows*, ed. K.C. Tsinganos, Kluwer, p. 459.

Khanna, R.,& Camenzind, M., 1996, *A&A*, **307**, 665.

MacGregor, K.B., 1996, in *Solar and Astrophysical Magnetohydrodynamical Flows*, ed. K.C. Tsinganos, Kluwer, p. 301.

Meier, D.L., Payne, D.G., & Lind, K., 1996, IAU Symp. **175**, 433.

Ray, T.P., Mundt, R., Dyson, J.E., Falle, S.A.E.G., & Raga, A.C., 1996, *Ap.J.Lett.*, **468**, L103.

Rüdiger, G., Elstner, D., & Stepinski, T.F., 1995, *A&A*, **298**, 934.

Schramm, K.-J., Borgeest, U., Camenzind, M. et al., 1993, *A&A*, **278**, 391.
Shu, F. et al., 1994, *Ap.J.*, **429**, 781.
Spindeldreher, S., & Appl, S., 1996, in *Proc. Heidelberg Workshop on Gamma-Ray Emitting AGN*, eds. J.G. Kirk, M. Camenzind, C. von Montigny, S. Wagner, MPI H - V37, p. 139.
Stapelfeldt, K.R. et al., 1995, *Ap.J.*, **449**, 888.
Wagner, S.J. et al. 1995, *A&A*, **298**, 688.

CHAPTER 2

RADIATIVELY DRIVEN JETS: A STATUS REPORT

L. NOBILI

Dipartimento di Fisica 'G.Galilei', Via Marzolo, 8, I-35131 Padova, Italy

The problem of the radiative acceleration mechanism is critically reviewed. The Compton drag proves the major obstacle for achieving relativistic velocities. However, in special cases the intense, dynamically comptonized radiation coming from the innermost regions of the source may lead to the formation of a two streaming configuration, possibly becoming the seed for the generation of jets.

1. INTRODUCTION

It is universally accepted that both gravitational and radiative forces are fundamental ingredients in modelling the central engine of compact galactic and extragalactic active sources. Much less obvious is the capability of the radiative forces in accelerating large amounts of matter up to relativistic velocities. In fact, despite their conceptual simplicity, all radiation–driven jets models derived so far failed in explaining either the high velocity ($\gamma \gg 1$) or the amazingly collimated features observed in a large number of peculiar astronomical objects. For this reason many theoreticians, struggling for the understanding of the physics of jets, addressed their efforts to the much more complex magnetohydrodynamical mechanisms based on the channeling of the rotational kinetic energy of the central 'engine' into the kinetic energy of the outflows (see Camenzind, Chapter 1). Unfortunately all these investigations assume adiabaticity and ignore possible effects arising from the interaction between the fast moving matter and the radiation field. This casts doubt upon the full validity of the models since a self–consistent hydrodynamical treatment, including photon propagation, meets with a series of difficulties that are common to those found in the genuine radiation–driven jet theory. In view of this situation, we can say that the question of what the accelerating mechanism really is, is still unanswered, despite the observational evidence that astrophysical jets are very common features at all scales.

2. THE DEVIL

Concerning the pure radiative mechanism, two levels of difficulty make life hard to theoreticians. The first one is related to the requisite that, in order to obtain a consistent transfer of momentum contrasting the intense gravitational attraction exerted by the central compact source, a very efficient, super-Eddington, radiator must be assumed to work in the system. The second problem, instead, is more properly related to some hydrodynamical properties of hot flows moving at a velocity near the velocity of light.

Actually, the strong limitation due to the need of assuming a super–Eddington flux is mitigated by postulating the existence of a very hot corona above the disc. For the time being, there is no profound physical reason in favor of this hypothesis, a part from the consideration that it works. In fact, if $T > 10^9$ K, the production of a conspicuous number of electron–positron pairs makes the gas lighter and, consequently, easier to be accelerated. In particular, in a two phase corona–disc model the minimum luminosity required to trigger a wind is reduced by a factor up to 10^{-3} when compared with the case of an electron–proton plasma. To describe more conveniently this situation I introduce here an adimensional luminosity ℓ^* in such a way that it coincides with the usual definition expressed in Eddington units, $\ell = L/L_{Edd}$, if the jet is made by ordinary electron–proton matter, but it is this quantity divided by $m_\Sigma/m_p \equiv m_e/m_p + n_p/(2n_e - n_p)$ if electron-positron pairs are involved, i.e. in general

$$\ell^* = \frac{m_p}{m_\Sigma} \frac{L}{L_{Edd}} \tag{1}$$

Here m_Σ is the *equivalent electronic mass* and n_p, n_e are the proton and total electron number density respectively. With this convention, the onset of a wind instability is fixed by the condition $\ell^* \geq 1$, no matter the composition of the fluid is. In particular, for a pure $e^\pm$ gas, the critical value $\ell^* = 1$ implies a luminosity which is $m_p/m_e \approx 2000$ times lower than the classical Eddington limit. I would like to point out, however, that the intense evaporation of the disc, and the consequent subtraction of a part of its internal energy, possibly prevent the existence of bodies having very large values of ℓ^*. Unfortunately, the feedback of the jet formation on the overall disc structure has been underestimated in the past, despite its relevance on the stability of these configurations.

Let us consider now more specifically the hydrodynamical problem. It was realized several years ago that it is not possible to accelerate optically thick flows up to relativistic velocities, contrary to the original, but incorrect believe that any exchange of energy and momentum between photons and particles occurs more efficiently when matter and radiation are strongly coupled. In particular, Turolla, Nobili and Calvani (1986) have shown that, taking into account the inertia of heat, the velocity of a radiation dominated, optically thick, spherically symmetric outflow is upper bounded by the value $v_{lim} = c/\sqrt{3}$. The existence of this limiting velocity is a consequence of the presence of a factor $\Delta_v = (1 - 3v^2)$ multiplying the luminosity term of the relativistic momentum equation (see Eq. 6 of their paper). Usually $v \ll c$, the coefficient Δ_v is nearly one, and radiation gives, all the way, a positive contribution to the acceleration of the fluid. However, when the velocity approaches the velocity of light, the coefficient Δ_v may change sign, leading to an apparent paradoxical result. In fact, if this condition is fulfilled, the radiative force works as a brake, rather than as an accelerator! Actually, this peculiar behaviour cannot surprise, since $c/\sqrt{3}$ is just the sound velocity of a relativistic isotropic fluid and, in our conditions, the photon dominated gas behaves exactly in this way (a quite similar inversion of the dynamical behaviour of a stationary flow during the critical sonic transition is a well known result of fluid mechanics). We note that the particular value of v_{lim} is related to the isotropy of the radiation field, as it results (indirectly) from an averaging of the velocity of light over the three independent space directions. This simple heuristic interpretation has an immediate physical consequence, because it suggests that high speed flows can be produced only in regions whose physical properties are dominated by 'well ordered', rather than chaotic relativistic fields. Regular magnetic configurations are special examples of these systems. Of course, in a radiation dominated region, the requirement of a well ordered field automatically implies that photons must propagate almost freely in a preferred direction, that is to say, the matter filling the region must be optically thin.

Unfortunately, even under such favorable conditions, a number of obstacles seem to conspire against the formation of high speed regimes. To illustrate this, let us consider the simple case of an optically thin flow propagating radially under the effect of an intense radiation coming from a finite region centered on the origin. For the sake of simplicity, in the following I will describe the motion using the ballistic approximation. Clearly, this is correct only if the fluid is supersonic and pressureless. On the other hand, the inclusion of a thermal gas component does not substantially modify our conclusions, although we are aware that, under very special conditions, the Compton rocket effect could play a role in favoring the acceleration (Cheng and O'Dell, 1981).

For a spherically symmetric gas distribution, assuming that the gravitational field is exerted by a point mass M centered on the origin, the momentum equation along the radial direction is

$$m_\Sigma \frac{\mathrm{d}\gamma\beta}{\mathrm{d}t} = \gamma^2 \sigma_T \left[(1+\beta^2)J - \beta(H+K)\right] - \frac{GMm_\Sigma}{r^2} \qquad (1)$$

where H, J and K are the first three moments of the radiation intensity (i.e. the flux, the energy density and the radiation pressure respectively), $\beta = v/c$, and σ_T is the Thomson cross section.

Assuming stationarity, with $d/dt = \gamma\beta c\, d/dr$, Eq. (1) can be recast into the non dimensional form

$$\gamma\beta\frac{\mathrm{d}\gamma\beta}{\mathrm{d}\ln\hat{r}} = \frac{1}{2\hat{r}} \left[\gamma^2(1+\beta^2)(\ell^* - 1) - \gamma^2\beta(2\frac{\ell^*}{\epsilon} - 1) - \frac{2\beta}{1+\beta}\right] \qquad (2)$$

where now $\hat{r} = c^2 r/(2GM) = r/r_g$ and ℓ^* is expressed in units of the generalized Eddington luminosity $L^*_{Edd} = m_p L_{Edd}/m_\Sigma$ defined above. The quantity

$$\epsilon = \frac{2J}{H+K} \qquad (3)$$

($0 \leq \epsilon \leq 1$) is related to the degree of collimation of the radiation field. For instance, for a radially streaming configuration, it is $\epsilon = 1$, while $\epsilon = 0$ for a fully isotropic radiation.

As an example, let us suppose that the radiation is emitted isotropically by a thin equatorial ring of matter with a radius $\hat{r}_D$, centered in the origin. This simple configuration mimics the presence of a geometrically thin disc because in these systems most of the emission is expected to arise from a ring a few tens r_g in radius. With this approximation, the expression of ϵ along the $\hat{z}$–axis turns out to be

$$\epsilon = \frac{2\hat{z}\sqrt{\hat{r}_D^2 + \hat{z}^2}}{\hat{r}_D^2 + \hat{z}^2} \quad \begin{cases} \approx 1 & \text{for} \quad \hat{z} \to \infty \\ \approx 0 & \text{for} \quad \hat{z} \ll \hat{r}_D \end{cases} \qquad (4)$$

Two important consequences can be deduced from Eq. (2). First, it is easy to verify that a significant acceleration occurs only very near the black hole horizon, that is in regions where the coefficient $1/\hat{r}$ is not too small. In fact, for radial distances much larger than the inner disc size (i.e. when $\hat{z} \gg \hat{r}_D > 1$) both gravitational and radiative forces become weaker and weaker, until their effects are completely overcome by all other perturbing forces arising from the interstellar medium (irregularities in the density distribution, shocks, magnetic fields, etc.). The second important information is obtained from the term in the squared brackets of (2). Equating to zero this quantity, we get the limiting value of the gamma Lorentz factor

$$\gamma_\infty \;=\; \frac{1}{\sqrt{2}} \left[1 + \frac{1}{\sqrt{1-\epsilon^2}} \right] \tag{5}$$

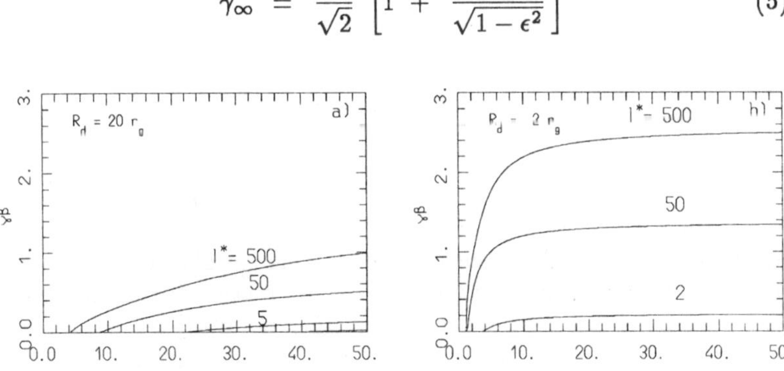

FIGURE 1. The run of the fluid velocity $\gamma\beta$ as a function of the radial distance for a radiatively driven flow. Radiation is assumed to be emitted by a thin ring of radius R_d (see text for details).

It is now evident that, since ϵ approaches to unity only at large distances, the braking action of the Compton drag, given by the second, negative term in Eq. (2), becomes increasingly larger at smaller distances. Consequently, we argue that the value of the terminal velocity critically depends on the competition of two opposite processes, but the most relevant aspect of the problem is that both these mechanisms enter in action only at distances comparable with the disc size.

This result can be seen more clearly by examining the graphs in Fig. 1a,b). The curves are obtained by integrating Eq. (2) for different values of ℓ^*, assuming $\hat{r}_D = 20$ (Fig. 1a) and $\hat{r}_D = 2$ (Fig. 1b). As we can see, the velocity of the outflows never reaches relativistic values unless ℓ^* is dramatically large, or the radius of the ring is comparable or smaller than $2\,r_g$ (!). Although the present derivation was obtained assuming a very special and oversimplified configuration, it is reasonable to expect that more complicated geometries do work in favor of an amplification of the Compton drag effect rather than of an increase of the outward radiative force. An example of an extreme situation is represented by a geometrically thick torus: inside the narrow funnels the radiation flux can be much larger than the Eddington limit, but the isotropy of the photons distributions, and the consequent Compton drag, prevent the formation of high speed fluid regimes.

2. THE ANGEL

In view of the situation described above, a tentative conclusion is that radiation is a quite inefficient mechanism for accelerating fluids. However, despite this pessimistic conclusion, there is still the possibility that radiative forces play a role in triggering collimated outflows of matter. The existence of this last beach is connected to a particular feature of electron–scattering dominated converging flows.

In fact, although we can expect that in the region between the inner edge of the disc and the central hole the gas distribution should be very chaotic, the intense gravitational attraction of the black hole forces the material to move mainly in the radial direction, so that, as a first approach, we can model the whole inner configuration with a free–falling spherical accretion. It is known that the radiative efficiency of the standard accretion is extremely low, since photons are created on a time scale much longer than the free–fall time scale. Indeed, even under the most favorable conditions and taking into account the most relevant emission mechanisms, the total luminosity of the spherical accretion turns out to be well below 10^{-5} Eddington units (Nobili, Turolla and Zampieri, 1991). However, a completely different situation occurs when the infalling gas is embedded in the radiation bath generated by the hot part of a luminous disc. To be crude, we can imagine that photons emitted by the disc, and subsequently diffused by the free electrons of the accreting gas, are created locally at the moment of the first scattering, as if a very efficient extra energy source should be located there. Part of this radiation is backscattered to infinity, contributing to increase the total luminosity detected by a distant observer. Clearly, the ratio L_{scatt}/L_{disc} between the luminosity of the inner portion of the accreting flow and the disc luminosity depends not only on the irradiation produced by the disc itself, but also on the fraction of diffused photons that succeed to escape to infinity. Actually, a consistent number of photons are dragged inward by the gas, being definitely lost once they cross the black hole horizon. The entire process is controlled by the location of the *trapping radius*, r_{tr}, defined as the distance where $\tau_{es}(r_{tr}) = 1/3$, being $\tau_{es}(r)$ the electron–scattering optical depth at r. Sometime r_{tr} is incorrectly considered as the surface below which photons cannot escape to infinity, being advected inward by the matter. However it is worthwhile to stress that, contrary to this widespread belief, the trapping radius does not work as a one–way membrane, since a number of photons produced near or below r_{tr} can still escape to infinity. In the presence of a strong advection, these photons are comparatively few (yet, the emergent flux is rather sensitive to the value of the optical depth), but the most interesting result is that just these photons are also the most strongly comptonized by the repeated scattering. As a consequence, the higher is the optical depth of the region where photons are created, the higher is their final energy (there is a sort of conspiracy in this process which replaces fews with quality!).

This phenomenon was studied early by Payne and Blandford (1981) and generalized to the relativistic case by Turolla et al. (1996). It is sometime referred to as *dynamical comptonization*, being germane to the more popular *thermal comptonization*. The relevance of the process depends on the value of the parameter $Y_d = \tau_{es}v$, which plays the same role of the thermal Compton parameter $Y_T = \tau_{es}(kT/m_ec^2)$. As in the last case, the transfer of energy is most efficient in electron scattering dominated region in which $Y_d \gtrsim 1$. Leaving the diffusive region, photons gain so much energy to dominate the high energy band of the spectrum. Neglecting relativity, the comptonized photons form a broad spectrum characterized by a power law, high energy tail whose spectral index is about 2. In normal conditions, i.e. for very low velocity ($v \ll 1$) and high optical thickness, the process is masked by true absorption which tends to thermalize matter and radiation, shaping the spectrum to the Wien distribution. For this reason, the effect of dynamical comptonization is particularly evident only when τ is not too large and the fluid velocity is high. Since near the horizon $v \lesssim 1$, even moderately values of the optical thickness produce a shift of the photon's energy up to the MeV; beyond this limit quantum effects succeed to steepen the spectrum. Moreover, Turolla et al. (1996) have shown that, when dynamical comptonization occurs near the black hole horizon, where relativistic aberration effects cannot be ignored, the energy distribution becomes even harder, with an index between 1 and 2 (see Fig. 2). This makes the entire process extremely interesting from an astrophysical point of view. In fact, the flattening of the high energy tail of the spectrum produce a sensible amplification of the radiation flux, causing the scattered luminosity L_{scatt} to be comparable, if not larger, than L_{disc}.

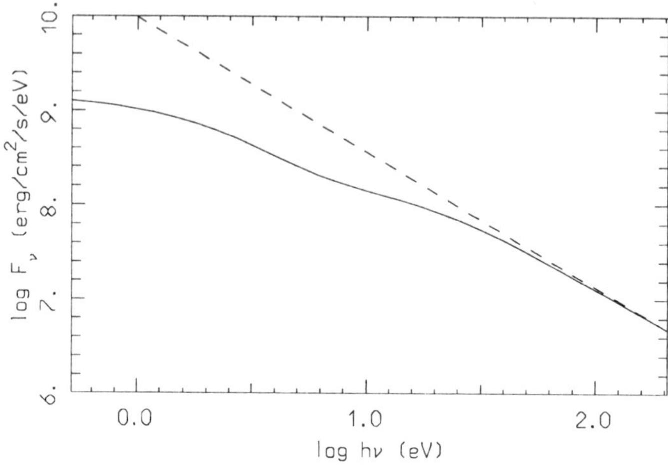

FIGURE 2. Emergent flux F computed using the CRM code (Zane et al., 1996) for a spherical accretion onto a Schwarzschild black hole. The derived spectral index is 1.36.

Dynamical comptonization can then provide a formidable machine for producing intense emission of X and γ-ray photons from the very inner regions of AGN. For this reason, it is also a promising candidate process for triggering outflows of neutral matter. In fact, the hard photons are originated inside a thin shell located near the horizon, they propagate radially, carrying out a large amount of energy and momentum. As discussed above, the small dimension of the source and the high energy flux are just the two fundamental ingredients needed to construct any relativistic jet model based on a radiative acceleration mechanism. Moreover, although the present results are obtained using a spherically symmetric environment, any deviation from this condition (for instance, an ellipsoidal rather than a spherical accretion) could help the formation of a two streaming configuration, in which the infalling matter is mostly concentrated near the equatorial plane while, under the effect of the intense γ and X-ray radiation coming from the inner region, a part of the gas outflows along the axes, as pictorially shown in Fig. 3. Indeed, the anisotropic distribution of the primary unscattered photons generated by the disc could contribute to increase further the anisotropy of the diffuse radiation. Finally, the particular streamline configuration of this picture automatically implies the presence of two annular stagnation regions located at some distance from the source. Then, it seems quite natural to associate these (presumably denser) zones with the BLR's observed in many active galaxies. In this hypothesis,

even the broadening and the blue-shifting of the atomic lines emitted by the clouds might be caused by the scattering of the reflected
radiation rather than by a Doppler shift effect.

I emphasize that dynamical comptonization is one of the most
thoroughly underestimated processes in astrophysics, despite that its
fantastic capability in boosting the photon energy may be the key to
the secrets of many astrophysical objects.

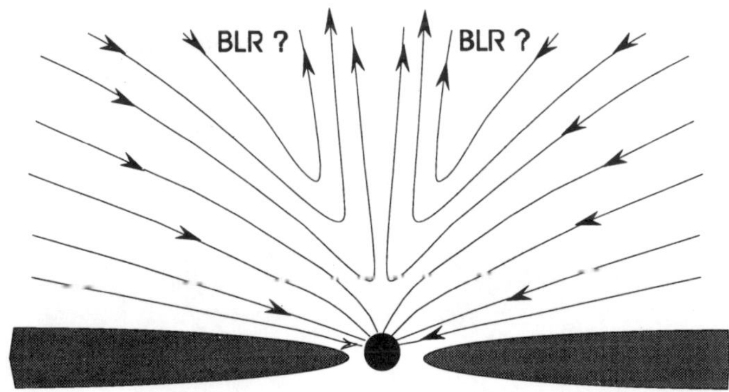

FIGURE 3. The figure illustrates schematically a possible scenario of the
inner region of an active source. Accretion occurs effectively only near the
equatorial plane. In the proximity of the axes the flow inverts its direction
under the push of the radiation emitted by the matter accreting into the
black hole.

3. CONCLUSIONS

At present, in the lack of any detailed calculations, the scenario
discussed in the above section must be considered only as a mere
working hypothesis, that needs to be proved (or disproved) using
a more correct approach. Just to carry out a test, a Monte–Carlo
simulation was applied to a simple model based on the hypothesis
that a spherically accreting material is illuminated by monochromatic
radiation emitted by a thin ring of radius $\hat{r}_D = 20$. In this simple
test, I made no attempt to include the general relativistic effects due
to the presence of the black hole. Trivially, photons are considered to
be lost when they cross the sphere $\hat{r} = 1$.

Assuming an accretion rate $\dot{m} = 10$ Eddington units, the calculated models produce the predicted sharp spectrum (see Fig. 4).

The amplification factor, derived computing the ratio between the total emitted energy and the energy detected far away, turns out to be $L_{scatt}/L_{ring} \approx 11$, but it can be larger, depending on the choice of the parameters $\dot{m}$ and $\hat{r}_D$. Unfortunately, due to the limitation of the code, no clear conclusion can be drawn about the anisotropy of the emerging flux. For a definite answer to this problem we need to solve the fully relativistic radiative transfer equations, assuming axial symmetry either for the gas distribution or for the photon direction, taking into account also for the correct Kline–Nishina cross section. To achieve this goal, the CRM method (Zane et al., 1996) seems to be a very promising strategy. Then the way forward is to extend the code to axisymmetric configurations.

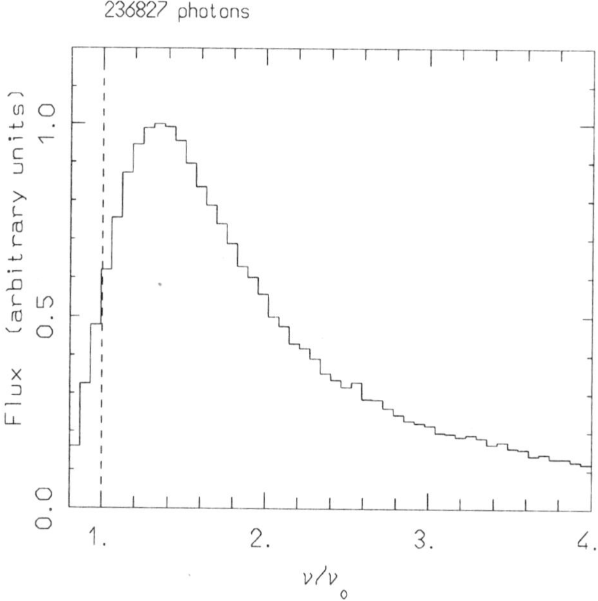

FIGURE 4. Emergent spectrum obtained from a Monte–Carlo simulation. A ring source located at $\hat{r} = 20$ is supposed to emit a monochromatic radiation at frequency ν_0. At infinity the spectrum is very broad and shifted to the blue.

REFERENCES

Cheng, A.Y.S. & O'Dell, S.L., 1981, *Ap.J.*, **251**, L49.
Nobili, L., Turolla, R., & Zampieri, L., 1991, *Ap.J.*, **383**, 250.
Payne, D.G. & Blandford, R.D., 1981, *MNRAS*, **196**, 781.
Turolla, R., Nobili, L., & Calvani, M., 1986, *Ap.J.*,. **303**, 573.
Turolla, R., Zane, S., Zampieri, L. & Nobili, L., 1996, *MNRAS*, **283**, 881.
Zane, S., Turolla, R., Nobili, L., & Erna, M., 1996, *Ap.J.*, **446**, 871.

CHAPTER 3

ON THE OSCILLATORY COLLIMATION OF ASTROPHYSICAL JETS

K. TSINGANOS AND N. VLAHAKIS

Department of Physics, University of Crete, GR-710 03
Heraklion, Crete, Greece

A notable common feature appearing in self-consistent, self-similar models of collimated MHD outflows, is that, before final cylindrical collimation is achieved, the jet passes from a stage of oscillations in its radius, flow speed and other physical variables. With rather general assumptions it is shown that this oscillatory behavior of magnetized outflows is not restricted to the few specific models studied so far, but instead it is a rather generic physical property of MHD outflows which start noncylindrically before they reach collimation.

1. INTRODUCTION

Previous studies have shown that under fairly general conditions magnetized outflows may become asymptotically cylindrical (Heyvaerts and Norman 1989). And, this tendency for asymptotic collimation has also been demonstrated via quasi-analytic self-similar solutions (Sauty and Tsinganos 1994, henceforth ST94, Contopoulos and Lovelace 1994, Trussoni et al. 1996). In all such self-similar solutions it is found that before final cylindrical collimation is achieved, the jet passes from a stage of oscillations in its radius, Alfvén number and other physical parameters.

A simple way to understand physically this effect is the following. Consider a single poloidal field/streamline $A(\varpi, z) = const.$ of an initially *radial* magnetized and rotating outflow that becomes asymptotically *cylindrical*. Assume for simplicity that the jet carries an electric current $I_z \propto \varpi^2$ with a uniform surface density $J_z = const.$ In its asymptotic regime the jet is confined by the interplay of the magnetic pinching force, the gas pressure gradient and the centrifugal force of rotation (ST94, Trussoni et al. 1996). Assume for simplicity that the gas pressure gradient is negligible such that at equilibrium the magnetic pinching force exactly balances the centrifugal force. In the superalfvenic regime, most of the conserved specific angular momentum is carried by the fluid, such that $L \approx \varpi V_\varphi$. The magnetic pinching force F_B which results from such a current I_z and the centrifugal force F_C for the assumed angular momentum conservation are then,

$$F_B = \frac{B_\varphi^2}{4\pi\varpi} + \frac{\partial}{\partial\varpi}\frac{B_\varphi^2}{8\pi} \propto \varpi, \qquad F_C = \frac{\rho V_\varphi^2}{\varpi} \propto \frac{L^2}{\varpi^3}, \qquad (1)$$

under uniform density conditions. If now at some equilibrium location 0, say at the cylindrical distance $\varpi = \varpi_0$, we have $F_B(\varpi_0) = F_C(\varpi_0)$, then at larger distances $\varpi_1 > \varpi_0$ we have according to Eq. (1) that $F_B(\varpi_1) > F_C(\varpi_1)$. Conversely, at the smaller cylindrical distances $\varpi_2 < \varpi_0$ we have again according to Eq. (1) $F_C(\varpi_2) > F_B(\varpi_2)$. The net result is that as the parcel of gas moves along the poloidal streamline from the central object to infinity, it feels an inward force at location 1 which brings it towards the rotation axis. On the other hand, due to inertia and its poloidal speed, it overpasses the equilibrium position 0 and arrives at location 2 where now feels an outward force bringing it again away from the rotation axis towards location 0, etc. The final result is the oscillatory shape of the streamline. The oscillations start at the collimation distance R_c where the streamlines start to deviate significantly from radiality and by means of the magnetic pinching forces are brought to the cylindrical geometry. Obviously, at large distances from the collimation radius R_c the cause of the oscillations disappears and accordingly their amplitude decays to zero, i.e., the uniform cylindrical shape is finally reached. An example of this situation has been given in Trussoni et al. (1996) where $A = f(R)\sin^2\theta$ with $f(R) \approx 1$ at $R \leq R_c$ while further away at $R \geq R_c$, $f(R) \approx R^2$. The poloidal magnetic field of such a case is a typical radial field $B_p \approx B_r = B_*/R^2 \cos\theta$ at $R \leq R_c$ while further away the poloidal magnetic field and flow become rather uniform similarly to the density at a given streamline. Besides the meridionally self-similar models (ST94, Trussoni et al. 1997), similar oscillations are found in radially self-similar (Blandford and Payne 1982, Fiege and Henriksen 1996), or translationally self-similar MHD models (Chan and Henriksen 1980, Bacciotti and Chiuderi 1992).

In the following we shall show under rather general assumptions that this oscillatory behavior of collimated outflows is not restricted to the few specific models studied so far, but instead it is a rather generic physical property of the MHD outflow as it reaches collimation. And, such topological stability of collimated outflows may be related to the Kelvin-Helmholtz (Ferrari et al. 1978, 1981, 1996, Bodo et al. 1994, 1995), or current driven instabilities (Appl and Camenzind 1992, Appl 1996).

2. PERTURBATIONS OF CYLINDRICAL OUTFLOWS

Consider an infinitely long jet where in a direction perpendicular to the flow axis, the outwards directed centrifugal force is balanced by the inwards tension of the toroidal magnetic field and gradient of the magnetic pressure, enhanced (reduced) by the gradient of the gas pressure,

$$\frac{\rho V_\phi^2}{\varpi} = \frac{\mathrm{d}}{\mathrm{d}\varpi}\left[\frac{B_\phi^2}{8\pi} + P\right] + \frac{B_\phi^2}{4\pi\varpi}, \tag{2}$$

where ϖ is the cylindrical distance in spherical coordinates (r, θ, ϕ), $\varpi = r\sin\theta$. In such a case of an asymptotically $(r \to \infty)$ collimated outflow (jet) (Heyvaerts and Norman 1989), the magnetic flux function A_∞, Alfvén number M_∞ and gas pressure P_∞ all become functions of the cylindrical distance ϖ at large radial distances r (in comparison to the Alfvén radius r_*) from the source of the outflow where we may neglect the gravitational field,

$$A_\infty = A_\infty\left(\varpi\right), \; M_\infty^2 = M_\infty^2\left(\varpi\right), \; P_\infty\left(\varpi\right) = \int \mathcal{F}_0 d\varpi, \tag{3}$$

with $\mathcal{F}_0(\varpi)$ some function of ϖ. For example, in the cases of cylindrical collimation of ST94 and Contopoulos and Lovelace (1994), $M_\infty\left(\varpi\right) = \text{const.}$ while $A_\infty\left(\varpi\right) \propto \varpi^2$, or, $A_\infty\left(\varpi\right) \propto \varpi^F$, for a constant F, in ST94 and Contopoulos and Lovelace 1994, respectively. In the following we shall check whether there exist small amplitude steady and axisymmetric perturbations in the streamline shape, Alfvén number and pressure, which satisfy the transfield MHD equations. We are interested to derive the dependence of these perturbations on the radial distance from the central object. Consider then a solution which is topologically close to one describing a collimated outflow,

$$A = A_\infty\left(\varpi\right)\left(1+\epsilon\right), M^2 = M_\infty^2\left(\varpi\right)\left(1+\epsilon_1\right), P = P_\infty\left(\varpi\right)+\delta P, \quad (4)$$

where all functions $\mid \epsilon \mid, \mid \epsilon_1 \mid$ and $\mid \delta P/P_\infty \mid \ll 1$. Substituting Eqs. (4) in the transfield equation (Tsinganos 1981) and by assuming that the derivatives of ϵ, ϵ_1 are also very small (so that we may ignore squares and products of the perturbation quantities), we obtain from the $\hat{z}$ and $\hat{\varpi}$ components of the momentum equation two equations which the perturbations ϵ, ϵ_1 and δP satisfy. Combining these two equations and assuming that the perturbations ϵ and ϵ_1 are linearly related, $\epsilon_1 = \lambda_0\left(\varpi\right)\epsilon$, we obtain a single equation for ϵ,

$$\frac{\partial^2 \epsilon}{\partial z^2} + \frac{\partial^2 \epsilon}{\partial \varpi^2} + \left[\frac{A'}{A}\left(2 - \frac{\lambda_0 M^2}{1-M^2}\right) - \frac{1}{\varpi}\right]_\infty \frac{\partial \epsilon}{\partial \varpi} +$$

$$\left[\mathcal{G}_1 + \lambda_0\mathcal{G}_2 - \lambda_0'\frac{M^2}{1-M^2}\frac{A'}{A}\right]_\infty \epsilon = 0, \quad (5)$$

where $\mathcal{G}_1$ and $\mathcal{G}_2$ are functions of ϖ (Vlahakis and Tsinganos 1997).

3. PERTURBATIONS SEPARABLE IN CYLINDRICAL AND SPHERICAL DISTANCE

The above equation relate the two unknown functions ϵ and λ_o and their derivatives. However, it is still complicated for a general analysis; in the following, we shall analyse Eq. (5) in the special case where the variables of the perturbations can be separated in the cylindrical and radial distances, (ϖ, r); separability of the coordinates in various other coordinates of the poloidal plane are analysed in Vlahakis and Tsinganos (1997). In the present case we assume,

$$\epsilon = f\left(\varpi\right)g\left(r\right), \qquad \mid g \mid \ll 1. \quad (6)$$

Then Eq. (5) gives,

$$g'' + \frac{g'}{r}\left\{2\varpi\frac{f'}{f} + \varpi\frac{A'}{A}\left(2 - \frac{\lambda_0 M^2}{1-M^2}\right)\right\}_\infty + g\left\{\frac{f''}{f} + \quad (7)\right.$$

$$\left.\left[\frac{A'}{A}\left(2 - \frac{\lambda_0 M^2}{1-M^2}\right) - \frac{1}{\varpi}\right]\frac{f'}{f} + \mathcal{G}_1 + \lambda_0\mathcal{G}_2 - \frac{\lambda_0' M^2}{1-M^2}\frac{A'}{A}\right\}_\infty = 0.$$

Therefore there are constants (s, k) such that:

$$g'' + 2s\frac{g'}{r} + k^2 g = 0,\tag{8}$$

or,

$$x^2\frac{d^2 y}{dx^2} + x\frac{dy}{dx} + \left[x^2 - \left(s - \frac{1}{2}\right)^2\right] y = 0, \quad x = kr, \quad y = gx^{s-\frac{1}{2}}.\tag{9}$$

The last differential equation is the familiar Bessel equation with the solution

$$y = D_1 J_{s-1/2}(x) + D_2 Y_{s-1/2}(x).\tag{10}$$

In the limit $x \to \infty$, Bessel's functions become,

$$J_\nu(x) \to \frac{1}{\sqrt{x}}\cos\left(x - \frac{\nu\pi}{2} - \frac{\pi}{4}\right), \quad Y_\nu(x) \to \frac{1}{\sqrt{x}}\sin\left(x - \frac{\nu\pi}{2} - \frac{\pi}{4}\right),\tag{11}$$

and therefore the solution is

$$g = \frac{D}{r^s}\sin\left(kr + \phi_0\right)\tag{12}$$

Finally, Eq. (7) gives two conditions relating the functions of ϖ:

$$\left\{2\varpi\frac{f'}{f} + \varpi\frac{A'}{A}\left(2 - \frac{\lambda_0 M^2}{1 - M^2}\right)\right\}_\infty = 2s\tag{13}$$

and

$$\left\{\frac{f''}{f} + \left[\frac{A'}{A}\left(2 - \frac{\lambda_0 M^2}{1 - M^2}\right) - \frac{1}{\varpi}\right]\frac{f'}{f} + \right.$$

$$\left. + \mathcal{G}_1 + \lambda_0 \mathcal{G}_2 - \frac{\lambda_0' M^2}{1 - M^2}\frac{A'}{A}\right\}_\infty = k^2.\tag{14}$$

In the following we shall apply and test our analysis in available analytical models of outflows which exhibit an oscillatory behavior.

4. EXAMPLES

We may start with the simplest case wherein M_∞, λ_0, f, $\mathcal{G}_1$, $\mathcal{G}_2$ are constants and $A_\infty(\varpi) \propto \varpi^2$. Indeed this case has been studied in

ST94 (see also Trussoni et al. 1996). They considered the following expressions of the free MHD integrals,

$$A = \frac{r_\star^2 B_\star}{2} \alpha(R, \theta), \quad \Psi_A(\alpha) = \frac{4\pi \rho_\star V_\star}{B_\star} \sqrt{1 + \delta\alpha}, \tag{15}$$

$$\alpha = \frac{\varpi^2}{r_\star^2 G^2(R)}, \quad R = \frac{r}{r_\star}, \tag{16}$$

$$L(\alpha) = \lambda r_\star V_\star \frac{\alpha}{\sqrt{1 + \delta\alpha}}, \qquad \Omega(\alpha) = \frac{\lambda V_\star}{r_\star} \frac{1}{\sqrt{1 + \delta\alpha}}, \tag{17}$$

where $G(R)$ is the radius of the jet in units of the Alfvén radius and λ, δ and $G(R \to \infty) = G_\infty$ are constants, while the starred quantities refer to values at the Alfvén radius $r_\star$. Writing down the expressions of the perturbations for this case we have,

$$\varepsilon(r) = \frac{G_\infty^2}{G^2(r)} - 1, \qquad \varepsilon_1(r) = \frac{M^2(r)}{M_\infty^2} - 1. \tag{18}$$

It follows from Eqs. (13 – 14) that $f(\varpi) = 1$ while λ_o is a constant which furthermore can be calculated at $r = r_*$,

$$\lambda_o \equiv -|\lambda_o| = -\frac{M_\infty^2 - 1}{M_\infty^2(G_\infty^2 - 1)}. \tag{19}$$

This is the same result with that in the study of ST94 although the surface $r = r_\star$ is not always in the asymptotic regime where gravity is negligible. The functions $\mathcal{G}_1$ and $\mathcal{G}_2$ are constants in this model and from Eqs. (13 – 14) the corresponding expressions for s and k are,

$$s = 2 + \frac{\lambda_0 M_\infty^2}{M_\infty^2 - 1} \tag{20}$$

$$k^2 \equiv \frac{4\pi^2}{\Lambda_{osc}^2} = \frac{2\lambda^2}{r_\star^2(1 - M_\infty^2)^2} \left[2 - \lambda_0 \frac{(2M_\infty^2 - 1) G_\infty^4 - M_\infty^4}{M_\infty^2(1 - M_\infty^2)} \right]. \tag{21}$$

The wavelength of the oscillations grows quadratically with the Alfvén number, $\Lambda_{osc} \propto M_\infty^2$, while the amplitude of the oscillations drops with distance as $r^{-(2-|\lambda_o|)}$, as found in ST94 (their Figs. 2, 8, 10). For example, as the magnitude of the asymptotic Alfvén number M_∞ increases by a factor of about 10 when the energetic parameter ϵ in the ST94 notation decreases from $\epsilon = 10$ to $\epsilon = 1$, Λ_{osc} increases accordingly by a factor of about 100. Similarly, the amplitude of the oscillations Λ_{osc} in the width of the jet and the Alfvén number drop with radial distance as r^{-s}, where $1 < s < 2$ with its exact value $s = 2 - |\lambda_o|$ depending on M_∞ and G_∞, according to Eq. (20).

Another more general class of solutions can be generated by the following set of free integrals,

$$A = \frac{r_\star^2 B_\star}{2}\alpha(R,\theta), \quad \Psi_A(\alpha) = \frac{4\pi\rho_\star V_\star}{B_\star}\sqrt{1 + \delta\alpha + \delta_\epsilon\alpha^\epsilon}, \qquad (22)$$

$$\alpha = \frac{\varpi^2}{r_\star^2 G(R)^2}$$

$$L(\alpha) = \lambda r_\star V_\star \alpha\sqrt{\frac{\alpha^{\epsilon-1} + \mu}{1 + \delta\alpha + \delta_\epsilon\alpha^\epsilon}}, \quad \Omega(\alpha) = \frac{\lambda V_\star}{r_\star}\sqrt{\frac{\alpha^{\epsilon-1} + \mu}{1 + \delta\alpha + \delta_\epsilon\alpha^\epsilon}}.$$

$$(23)$$

where μ, ϵ, δ_ϵ are constants, in addition to the ones introduced in the previous example.

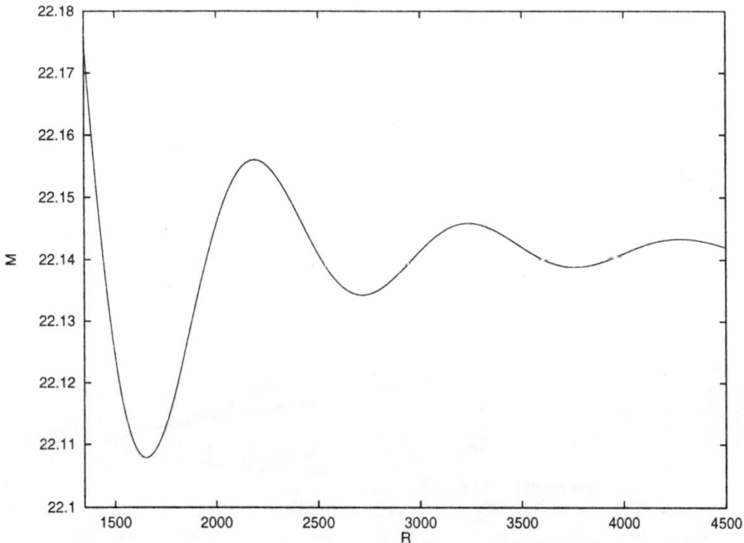

FIGURE 1. Alfven number M vs. radial distance R for the model of example 2 and for the following set of parameters: $\epsilon = 0.1$, $\delta = 0.35$, $\delta_\epsilon = 10^{-4}$, $\lambda = 0.1$, $\mu = 500$. We find from the integration of the transfield MHD equation that $M_\infty^2 = 490.24$, $G_\infty^2 = 0.0769$.

If M_∞, λ_0 and f are constants then,

$$k^2 = \frac{2\mu\lambda^2(\epsilon - 1)(G_\infty^4 - M_\infty^2)}{r_\star^2 M_\infty^2(1 - M_\infty^2)^2} \approx \frac{2\mu\lambda^2(1 - \epsilon)}{r_\star^2 M_\infty^4} \qquad (24)$$

because $M_\infty^2 \gg 1$,

$$\lambda_0 = \left[(\epsilon + 1)\,M_\infty^2 - (\epsilon - 1)\,G_\infty^4\right] \frac{1 - M_\infty^2}{\left(2M_\infty^2 - 1\right)G_\infty^4 - M_\infty^4}\,, \qquad (25)$$

and

$$s = 2 + \frac{\lambda_0 M_\infty^2}{M_\infty^2 - 1} \approx \epsilon + 3\,. \qquad (26)$$

Substituting in Eq. (4) the above expressions for k and s we find the perturbed form of the stream function,

$$A \approx \frac{B_\star \varpi^2}{2G_\infty^2} \left[1 + \frac{D_0}{R^{\epsilon+3}} \sin\left(\frac{\sqrt{2\mu\lambda^2\,(1 - \epsilon)}}{M_\infty^2} R + \phi_0\right)\right]\,. \qquad (27)$$

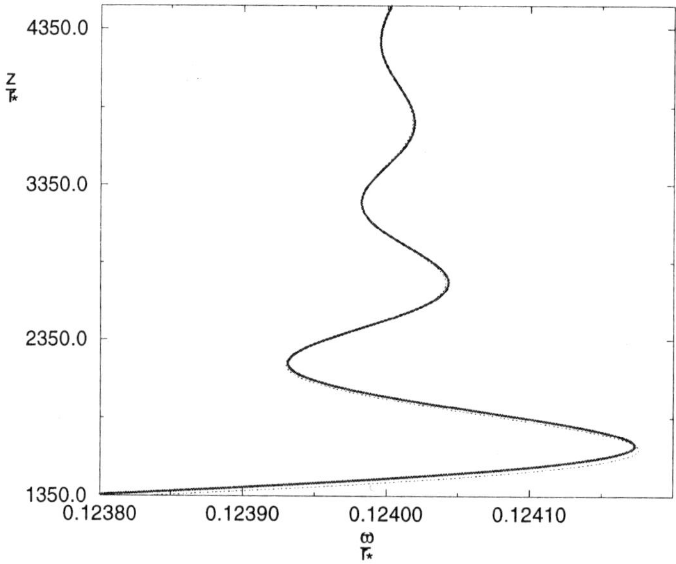

FIGURE 2. Shape of the streamline on the poloidal plane for the model of example 2, as emerges from the perturbation analysis ($D_0 = 2.79 \times 10^7, \phi_0 = 0.46$). With dotted line an exact solution is shown for the model of example 2 where gravity is included.

Acknowledgements

This research has been supported in part by a grant from the General Secretariat of Research and Technology of Greece. We thank E. Trussoni, C. Sauty and G. Surlantzis for helpful discussions.

REFERENCES

Appl, S., & Camenzind, M., 1992, *A&A*, **256**, 354.

Appl, S., 1996, *A&A*, **314**, 995.

Bacciotti, F., & Chiuderi, C., 1992, *Phys. Fluids*, 4, 35.

Bodo, G., Massaglia, S., Ferrari, A., & Trussoni, E., 1994, *A&A*, **283**, 655.

Bodo, G., Massaglia, S., Rossi, P., Rosner, R., Malagoli, A., & Ferrari, A., 1995, *A&A*, **303**, 281.

Blandford, R.D., & Payne, D.G., 1982, *MNRAS*, **199**, 883.

Chan, K.L., & Henriksen, R.N., 1980, *Ap.J.*, **241**, 534.

Contopoulos, J., & Lovelace, R.V.E., 1994, *Ap.J.*, **429**, 139.

Ferrari, A., Trussoni, E., & Zaninetti, L., 1978, *A&A*, **64**, 43.

Ferrari, A., Trussoni, E., & Zaninetti, L., 1981, *MNRAS*, **196**, 1051.

Ferrari, A., Massaglia, S., Bodo, G., & Rossi, P., 1996, in *Solar and Astrophysical MHD Flows*, K. Tsinganos (ed.), Kluwer Academic Publishers, 607.

Fiege, J.D., & Henriksen, R.N., 1996, *MNRAS*, **281**, 1038.

Heyvaerts, J., & Norman C.A., 1989, *Ap.J.*, **347**, 1055.

Sakurai, T., 1987, *PASJ*, **39**, 821.

Sauty, C., & Tsinganos, K., 1994, *A&A*, **287**, 893 (ST94).

Trussoni, E., Sauty, C., & Tsinganos, K., 1996, in *Solar and Astrophysical MHD Flows*, K. Tsinganos (ed.), Kluwer Academic Publishers, 383.

Tsinganos K., 1982, *Ap.J.*, **252**, 775.

Tsinganos, K., Sauty, C., Surlantzis, G., & Trussoni, E., Contopoulos J., 1996, in *Solar and Astrophysical MHD Flows*, K. Tsinganos (ed.), Kluwer Academic Publishers, 247.

Vlahakis, N., & Tsinganos, K., 1997, *MNRAS*, in press.

CHAPTER 4

COLLIMATION CRITERIA IN STELLAR OUTFLOWS : FROM JETS TO WINDS

C. SAUTY[1], E. TRUSSONI[2], K. TSINGANOS[3]

[1]*Observatoire de Paris - Université Paris 7, DAEC URA 173, F-92190 Meudon, France*

[2]*Osservatorio Astronomico di Torino, Strada dell'Osservatorio 20, I-10025 Pino Torinese, Italy*

[3]*University of Crete, Dept. of Physics, GR-71003 Heraklion, Greece*

By modeling stationary, rotating and magnetized outflows, self–similar in the meridional direction, we show that magneto–centrifugally driven winds can either collimate into thin jets or expand into wider bipolar flows, depending on the energy input into the flow and on the pressure distribution across the jet. The asymptotic analysis shows that overpressured winds, with respect to their environment, are radially expanding unless the central object is a Fast Magnetic Rotator (FMR), in which case the strong magnetic pinch collimates the outflow. Conversely underpressured winds are always cylindrically collimated. However they can be either magnetically focalized in a thin jet, as in the case of FMR, or less efficiently confined by the external pressure, appearing as bipolar flows in the case of Slow Magnetic Rotators (SMR). We can draw out a scenario for the evolution of cylindrically collimated jets into winds as the central star evolves from FMR to SMR, and wonder whether the solar wind itself is indeed a collimated bipolar outflow on large scales.

1. INTRODUCTION

The large scale ejection of magnetized plasma from astrophysical objects is observed either in the form of winds, with a more or less spherical or bipolar shape, or in the form of tightly collimated jets. Winds and bipolar flows are observed around evolved stars like the Sun, early–type O/B stars and planetary nebulae. On the other hand, jets are observed in association with accretion discs around Young Stellar Objects (YSOs) and galactic compact sources. A comparable dichotomy may also exist in extragalactic sources where well focalized jets from strong radio sources like FRII are observed, in contrast to less powerful bipolar flows from Seyfert galaxies.

One question to be addressed is whether the formation of collimated jets is strictly connected to the presence of an accretion disc (Blandford and Payne 1982) or to the nature itself of the rotating magnetized plasma outflowing from the central source. According to the analysis of Heyvaerts and Norman (1989) all polytropic winds, regardless of their source, should collimate (or at least part of them) into a cylindrical jet. However, the radius of the jet and the distance from the source at which they collimate may vary critically with the parameters. This severely constrains their appearance either as a jet or just as a bipolar flow. In turns, it may be the presence of the disc that provides the boundary conditions to the collimated wind to appear as a jet.

In order to get some insight on this issue of the collimation process, one can concentrate on getting fully consistent axisymmetric solutions of the MHD equations describing a steady outflow from a central gravitating body, while releasing the polytropic assumption. This can be done by means of a meridional self–similar model as developed by Tsinganos et al. either prescribing the geometry of the flow (Trussoni et al. 1997, henceforth TTS97) or assuming the latitudinal dependence of the pressure and deducing the geometry of the streamlines (Sauty and Tsinganos 1994, henceforth ST94). This self–similar approach allows to describe the outflow structure close to the rotational axis.

In the first case (prescribed pattern of the streamlines, TTS97) it has been shown that a wind attains cylindrical collimation *only* via the combination of the rotation and the toroidal component of the magnetic field (Trussoni and Tsinganos 1993). The asymptotic equilibrium is magnetically dominated in the case of Fast Magnetic Rotators (FMR) but it may be either magnetically or pressure dominated in Slow Magnetic Rotators (SMR). Pressure dominated jets are of course underpressured (i.e. the pressure increases by receding away from the axis) in order to keep the confinement, although magnetic stresses are essential to reach this collimated configuration. They also achieve larger terminal velocities than purely magnetically confined, overpressured jets (i.e. with the pressure decreasing receding away from the axis).

Following the second approach (free streamlines), it has been found that winds with a spherically symmetric pressure distribution collimate either in the form of a cylindrical jet or expand radially (ST94). In such a case the outflow is always magnetically dominated asymptotically, and its behavior depends critically on the angular distribution of energy input and losses at the base. Here we extend these results by assuming a non spherically symmetric structure for the pressure. We will not discuss yet the complete analysis of this problem, that requires the solution of a system of three stiff ordinary differential equations (ST94, Tsinganos et al. 1996). We present conversely, following an asymptotic analysis, some new results on the collimation criterion, that depends whether the jets are under- or overpressured, and in magnetically or pressure dominated regimes.

2. PARAMETERS OF THE MODEL

In steady and axisymmetric ideal MHD the following free integrals exist, that are constant along each poloidal streamline: the magnetic flux function A, the mass flux to magnetic flux ratio $\Psi_A(A)$, the total angular momentum $L(A)$ and finally the angular velocity of the line footpoints $\Omega(A)$. Note that the quantity ΩL represents the energy of the magnetic rotator, i.e. the energy necessary to bring the field lines in rotation in the comoving frame of the fluid (Belcher and MacGregor 1976). The poloidal velocity of the flow equals the poloidal Alfvén speed at the cylindrical radius $\varpi_a = \sqrt{L/\Omega}$. We assume also that surfaces of constant poloidal Alfvén Mach number are spherical, $M_a^2 = M_a^2(r)$, while the magnetic flux function is expanded with the colatitude θ to the first dipolar term, $A = A_\star f(r) \sin^2 \theta$. The free integrals are chosen accordingly: $\Psi_A^2 = 4\pi\rho_\star[1 + \delta(A/A_\star)]$, $L\Psi_A = r_\star B_\star \lambda(A/A_\star)$ and $\varpi_a^2 = r_\star^2(A/A_\star)$. To satisfy Euler's equation, the pressure distribution is $P = \frac{1}{2}\rho_\star V_\star^2 Q(r)(1 + \kappa A/A_\star)$. The quantities $r_\star$, $B_\star$, $V_\star$ and $\rho_\star$ are values taken along the polar axis at the Alfvén transition where $M_a^2 = 1$, and $A_\star$ is simply $r_\star^2 B_\star/2$.

The variation with the colatitude of each physical quantity in the flow then depends on a single parameter. The increase or decrease of the density with the colatitude is measured by δ ($\delta > 0$ means a plasma denser at the equator than at the pole, at the same radial distance). The behavior of the pressure is ruled by κ: for $\kappa > 0$ or $\kappa < 0$ the flow is underpressured or overpressured, respectively. The strength of the energy of the magnetic rotator depends on λ^2. If the flow is cylindrically collimated, the asymptotic value of the pressure Q_∞ is assumed as a new parameter, that results from the choice of the initial pressure at the base of the flow.

The last parameter of the model is related to the energetics of the wind. Denote by E the total energy density per unit mass along each streamline:

$$E \equiv \frac{1}{2}V_{\rm p}^2 + \frac{1}{2}V_\varphi^2 + h - \frac{\mathcal{G}\mathcal{M}}{r} - \frac{\Omega}{\Psi_A}r\sin\theta B_\varphi - \int_{s_o}^{s}\frac{q}{\rho V_{\rm p}}{\rm d}s\,, \qquad (1)$$

where $\mathcal{G}$ is the gravitational constant, $\mathcal{M}$ the mass of the star, h the enthalpy of the gas, s the path length along a field line A, and q the local heating rate. Despite the fact that the present model is not polytropic, it turns out that the variation with colatitude of the energy per unit volume is simply proportional to the magnetic flux. This can be parameterized by a single quantity ϵ related to the amount of excess energy available at any nonpolar streamline as compared to the polar one. Normalizing this variation of energy with the volumetric energy of the magnetic rotator, we obtain finally (ST94):

$$\frac{\epsilon}{2\lambda^2} \approx \frac{\rho E - \rho_{\rm pole}E_{\rm pole}}{\rho L\Omega} \approx 1 - \frac{E_G}{\rho\Omega L}\,. \qquad (2a)$$

$$E_G = (\rho - \rho_{\rm pole})\frac{\mathcal{G}\mathcal{M}}{r}\,, \qquad (2b)$$

Thus, the sign of this quantity depends mainly on the difference of the energy per unit volume of the magnetic rotator $\rho\Omega L$ and the colatitudinal variation of the gravitational energy per unit volume E_G. The faster and the more magnetized is the source (the higher is λ) the more positive ϵ/λ^2 tends to be. On the opposite, the higher is the excess of density at the equator as compared to the pole (the higher is $\delta > 0$), the more negative ϵ/λ^2 tends to be. In conclusion, the solutions depend on δ, κ/λ^2, Q_∞, λ^2 and ϵ/λ^2

3. RESULTS

An analytical study of the force balance can be carried out in the asymptotic region $(r \gg r_\star)$ for a cylindrically collimated flow. In this case the dynamical equilibrium across the streamlines is $(R = r/r_\star)$:

$$\rho\frac{V_\varphi^2}{R} = \frac{d}{dR}\left(\frac{B_\varphi^2}{8\pi}\right) + \frac{B_\varphi^2}{4\pi R} + \frac{\partial P}{\partial R}\,, \qquad (3)$$

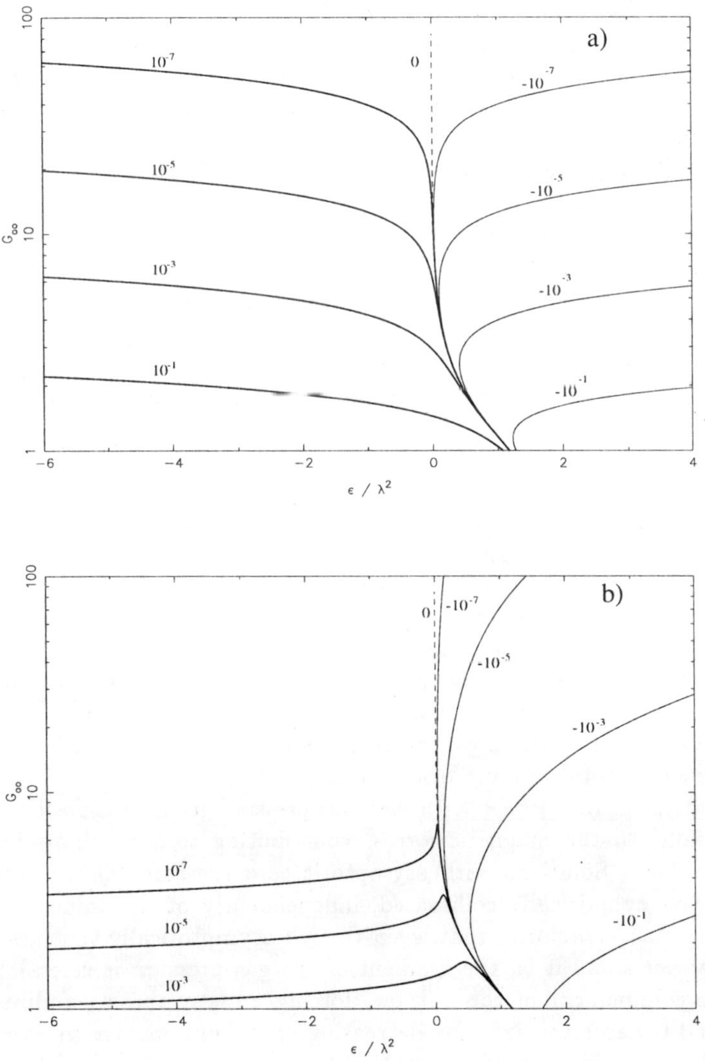

Fig. 1. Plot of the asymptotic cylindrical radius G_∞, in units of the Alfvén radius, versus ϵ/λ^2 for vanishing ($Q_\infty = 0$, panel a) and finite ($Q_\infty = 10$, panel b) asymptotic pressure. Labels indicate different positive (thick solid lines) and negative (thin solid lines) values of κ/λ^2 (on the dashed line $\kappa/\lambda^2 = 0$).

where the centrifugal force prevents the magnetic forces from pinching the flow on its axis. The pressure gradient ($\propto \kappa Q$) can contribute to or prevent the collimation depending on the sign of κ. Combining Eqs. (2) and (3) with our assumptions, we get the asymptotic values of the Alfvén number, of the cylindrical radius of the jet and of the velocity, normalized to their value at the Alfvén transition. In Fig. 1 we plot the non dimensional radius of the jet G_∞ vs ϵ/λ^2: the bigger is G_∞, the wider is the jet's lateral extension and the less strongly collimated is the outflow.

Equally pressured jet ($\kappa = 0$): It was found (ST94) that full collimation is reached only for $\epsilon/\lambda^2 > 0$, which corresponds to FMR. Conversely, when $\epsilon/\lambda^2 \leq 0$ the wind does not collimate and remains radially expanding, with $G_\infty \to \infty$ (Fig. 1). In the former case the Alfvén number attains a finite value, while it diverges in the wind with radial streamlines.

Overpressured jet ($\kappa < 0$): The pressure gradient is negative i.e. it is opposite to the magnetic pinching. We can see (Fig. 1) that collimated solutions exist only for $\epsilon/\lambda^2 > (\epsilon/\lambda^2)_{\text{lim}} > 0$, with the value of $(\epsilon/\lambda^2)_{\text{lim}}$ depending on κ/λ^2. If ϵ/λ^2 is lower than that limit, the numerical analysis shows that radial solutions may exist as in the case with $\kappa/\lambda^2 = 0$. Accordingly, the criterion for collimation is not changed for overpressured jet, as the limiting value $(\epsilon/\lambda^2)_{\text{lim}}$ is close to 0. For $\epsilon/\lambda^2 > (\epsilon/\lambda^2)_{\text{lim}} > 0$ there are two branches of collimated solutions. The lower one, close to the branch $\kappa/\lambda^2 = 0$, can fulfill the stellar boundary conditions smoothly crossing the Alfvén singularity, as tested in the numerical solutions. As the value of $(\epsilon/\lambda^2)_{\text{lim}}$ is not strongly affected by Q_∞ (compare Figs. 1a and 1b), it is evident that the main structure of overpressured jets is not critically affected by the asymptotic pressure of the gas.

Underpressured jet ($\kappa > 0$): Now the pressure gradient is positive and parallel to the magnetic forces, contributing to the collimation of the flow. Solutions with asymptotic zero pressure ($Q_\infty = 0$) are always cylindrically collimated, independently of the value of ϵ/λ^2 (Fig. 1a). We notice that, even though asymptotically vanishes and however small it is, the gradient of the gas pressure is essential for the confinement of the outflow, forcing anyway the streamlines to bend towards the axis. By decreasing ϵ/λ^2 from positive to negative values for a given κ/λ^2, we see that the beam radius rapidly increases, reaching then a plateau in the region of negative ϵ/λ^2. Qualitatively we may say then that the criterion for collimation is conserved: for positive ϵ/λ^2 the outflow appears very well collimated with low values of G_∞. Conversely, for negative ϵ/λ^2 the asymptotic radius of the beam is much larger and it could appear as a bipolar flow.

The effect of a non vanishing asymptotic pressure ($Q_\infty > 0$)

is shown in Fig. 1b: for decreasing values of ϵ/λ^2 the jet radius increases, but it reaches a maximum and then lowers (the velocity conversely increases monotonically). In the region where G_∞ grows up, the magnetic field is mainly responsible for the confinement, as in the case with $Q_\infty = 0$. At the maximum asymptotic radius the pressure gradient and the magnetic stress are equally balancing the centrifugal force, in a sort of equipartition state. On the left of the maximum, G_∞ decreases because the asymptotic pressure gradient is becoming too important, and we may say that the jet is confined by the gas pressure.

We may conclude that overpressured jets can be confined only by magnetic forces, but this happens only in the case of FMR ($\epsilon/\lambda^2 > 0$). On the opposite, underpressured winds are confined cylindrically by the magnetic forces in the case of FMR and by the pressure gradient in the case of SMR. In this last case however the confinement is less efficient and the outflow appears as a bipolar flow rather than a jet.

3. DISCUSSION

From these results we may try to outline some possible scenario for the evolution of jets into winds. We will concentrate on the case $\kappa/\lambda^2 > 0$ that seems observationally favored, at least for the case of solar coronal holes and some YSOs. As a star evolves, it looses angular momentum decreasing thus the available energy of the magnetic rotator: accordingly we expect that ϵ/λ^2 changes from positive to negative values. Furthermore we can also expect that as the jet gets wider, the pressure gradient and its asymptotic value decrease. Therefore the outflow evolution cannot be simply given on Fig. 1b by moving in the direction of reducing ϵ/λ^2, but also by jumping from Fig. 1b to Fig. 1a. This effectively increases further the asymptotic radius of the jet. We may suggest that the evolution tries to follow a configuration of maximum radius such that the pressure gradient and the magnetic pinching are always equal to each other. As the star evolves from FMR to SMR, the jet would continuously get wider, until it becomes a bipolar flow with a vanishing asymptotic pressure and a large radius. If this scenario is reasonable, it implies also that the solar wind could be asymptotically collimated, though it would not quite appear as a jet.

REFERENCES

Belcher, J.W. & McGregor, K.B., 1976, *Ap.J.*, **210**, 498.

Blandford, R.D. & Payne, D.G., 1982, *M.N.R.A.S.*, **199**, 883.

Heyvaerts, J. & Norman, C.A., 1989, *Ap.J.*, **347**, 1055.

Sauty, C. & Tsinganos, K., 1994, *A&A*, **287**, 893 (ST94).

Trussoni, E. & Tsinganos, K., 1993, *A&A*, **269**, 589.

Trussoni, E., Tsinganos, K. & Sauty, C., 1997, *A&A*, (in press) (TTS97).

Tsinganos, K., Sauty, C., Trussoni, E., Surlantzis, G. & Contopoulos, J., 1996, *M.N.R.A.S.*, **283**, 811.

CHAPTER 5

JETS FROM STARS AND GALACTIC NUCLEI

W. KUNDT

Institut tur Astrophysik der Universität Bonn,
Auf dem Hügel 71, D–53121 Bonn, Germany

Various mechanisms – and central engines – have been proposed throughout the past thirty years to explain the various observed jet sources. My own preference is in favor of rapidly rotating magnets generating relativistic pair plasma in localized magnetospheric discharges. The centrifugally pressurized (magnetized) pair plasma behaves as a weightless, inmiscible fluid which tries to escape from the (bottom of the) potential well like air bubbles escape from a bath tub, thus ramming two antipodal vacuum channels along the route of least resistance, in the vicinity of the spin axis of the feeding disc. In the case of the extragalactic jet sources (from the rotation centers of 'active' galaxies), inverse-Compton losses (on the ambient photon bath) limit the escape probability to $\gtrsim 10\%$ – the fraction of 'radio-loud' sources – whereas same probability appears to be near 100% in the cases of jets from newborn stars, young binary white dwarfs, and young binary neutron stars. Once beyond the central high-pressure, subsonic 'broad-line region', the relativistic plasma escapes in the form of two cold, supersonic, antipodal jets.

1. THE BINARY-FLOW FAMILY

The family of astrophysical jet sources – or bipolar flows – involves four different classes of central engines: (1) the rotation centers of galaxies, (2) young stellar objects (YSOs, or pre-T-Tauri stars), (3) young binary neutron stars (like SS433), and (4) young binary white dwarfs, the central lighthouses of planetary nebulae (cf. Kundt 1996b, and Fig. 1). Their common characteristics are (i) *elongated* morphologies (lobes, cocoons), containing (ii) *narrow jet channels* (opening angle of order 10^{-2}) which involve emission *knots* (Herbig-Haro objects, FLIERS), *hotspots* and/or heads (bow shocks), and which are quite often *one-sided*, due either to relativistic beaming, or to enhanced absorption, (iii) *lobe/core power ratios* of $10^{-2\pm2}$, (iv) rapid *core variability*, (v) very *broad core spectra*, from radio to γ-ray frequencies, or even VHEs (in classes (1) and (3)), and often (vi) *superluminal motion* of knots in their jets.

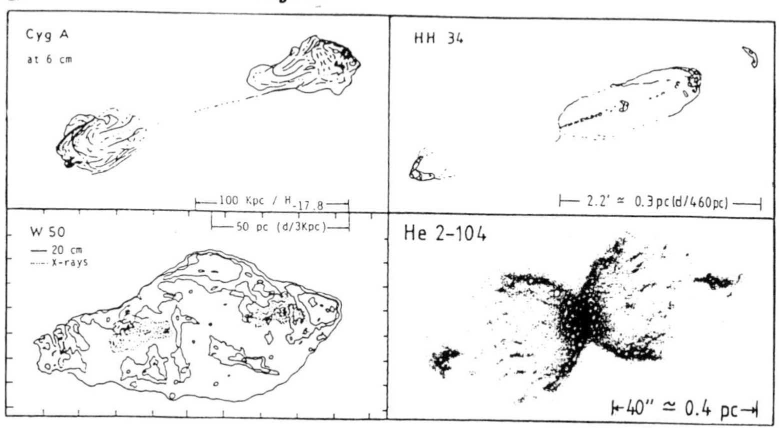

FIGURE 1. Sketch of representatives of the *Bipolar-Flow family*, one for each class. The right-hand column contains optical maps, predominantly from emission lines. SS 433, the representative of neutron-star driven jets, is confused by the simultaneous presence of the supernova shell W 50 which is thought to owe its existence to the birth of the neutron star.

The properties (i) to (vi) of the bipolar-flow family are not all obvious in all cases. A detection of *narrow jets*, and of their knots requires high resolution, and depends on the amount of flow resistance, i.e. on the circumstellar medium: an unperturbed, cold, relativistic jet performs a loss-free $\vec{E} \times \vec{B}$-drift, hence is invisible. The stormier its environment, the brighter a jet. This property is alien to most numerical simulations which assume (warm and/or turbulent) magnetohydrodynamic motion.

Next consider *sidedness*. Sidedness of the central jets is not only a common property of the extragalactic sources but is likewise encountered in the Galactic binary SS433, the Galactic superluminals GRS 1915+105 and GRO J1655-40, and the soft γ-ray repeater 1806-20, all of which are likely to be *neutron-star* binaries (Kundt 1991, 1996b). (In my opinion, the BH-candidates like GRS 1915+105 are neutron stars surrounded by massive accretion discs: Kundt 1996c, 1997). Sidedness is not a familiar property of *planetary nebulae*, but see M1-92 mapped by Trammell and Goodrich (1996)! Finally, (non-absorptive) sidedness has been encountered in (the nuclear radio jets of) YSO sources such as S 68 in Serpens, HH 1-2, and HH 111: Reipurth and Heathcote (1993), Rodríguez and Reipurth (1994); cf. Fig. 2a,b). Such observations rule directly against non-relativistic bulk speeds in the YSO class; cf. Blome and Kundt (1988) where relativistic jet speeds are derived from basic assumptions. A more indirect evidence against focused stellar winds – the present-day doctrine – are the large velocity asymmetries (factor of 2 ± 0.5) observed in 50% of all YSO sources: Hirth, Mundt and Solf (1994).

A key representative of the neutron-star driven jet sources is the Galactic binary SS433. It takes a lot of room in my list of publications, among them slots in 1991 and 1996b (p.140). In my understanding, its multiple blue- and red-shifted optical and X-ray emission lines are not emitted by 'bullets' of local galactic material shot out along the straight-line generators of a precessing cone but by material at the inner edge of its non-accretion disc, in quasi-Keplerian motion. (The standard model for SS433 is multiply inconsistent). The jet substance in SS433 is like that of all the other jet sources: relativistic pair plasma in supersonic motion through vacuum channels dragged around by the orbital motion, and redirected by a precessing injection direction. The jets approach us in the West where they are much brighter than in the East. Note that similar (velocity-broadened) high-Temperature X-ray emission lines (from Fe- and other high-Z-ions) have been recently reported – not only from Seyfert galaxies, but also – from the Cornet protostar cluster in the R CrA molecular cloud: Koyama et al. (1996).

Returning to properties (iii) to (vi) above, we understand a (bolometric) *lobe/core power ratio* of 1% as a 1% average *efficiency* of the central engine to blow jets, and an ensemble fluctuation by $10^{\pm2}$ around this value as due to the core's short-time variability (compared with the much smaller time-integrated variability of the lobes which are blown by the jets). The *broad core spectra* tend to be understood as superpositions of thermal, synchrotron, and inverse-Compton radiation whereby naked-core extragalactic sources can have their spectral power peak above 1TeV (visible for sufficiently nearby sources), more than 10 orders of magnitude above the (thermal) Big Blue Bump (BBB), corresponding to average *Lorentz factors* above 10^5 of the colliding electrons (and positrons): AGN electron spectra have average Lorentz factors above 10^5. The corresponding estimate for YSO sources, with equipartition-synchrotron Lorentz factors of order 10^3 and cooler thermal BBBs (than 10^6 K), leads to spectra extending

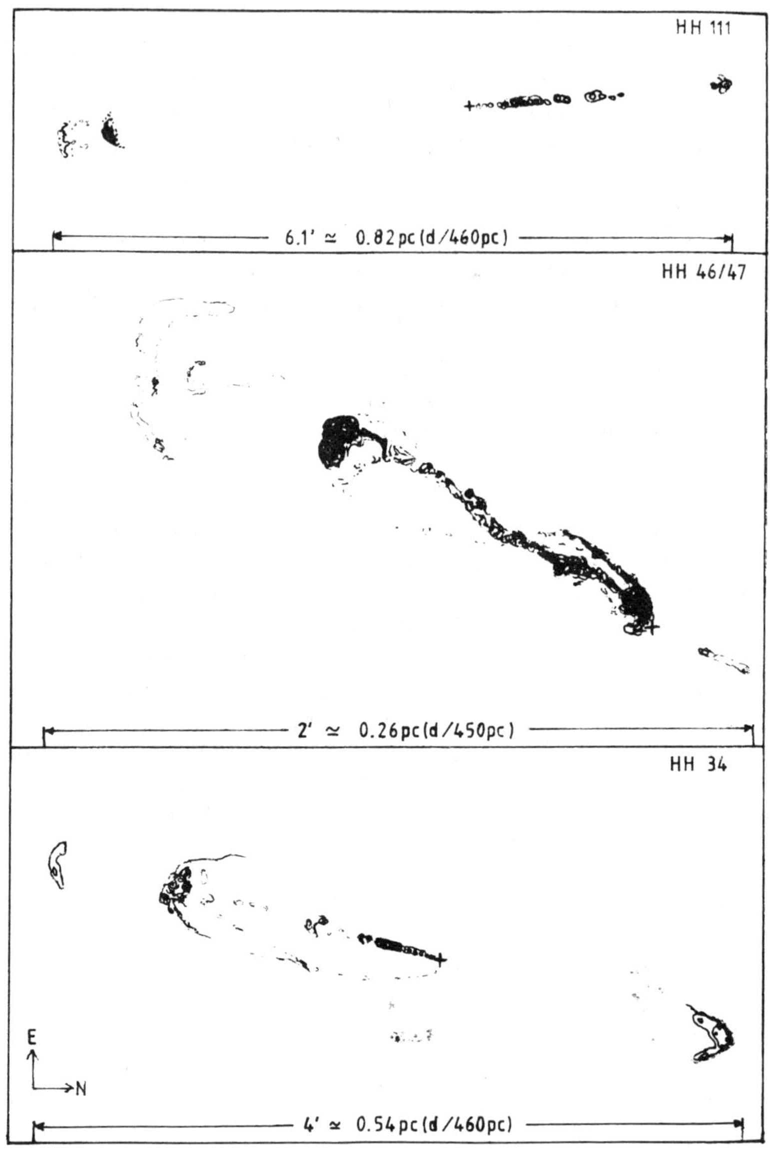

FIGURE 2a. Three well-mapped *YSO jet sources*, at Hα and [S II], taken from Kundt (1996d). It is my (often argued) conviction that all of these sources involve pair-plasma jets.

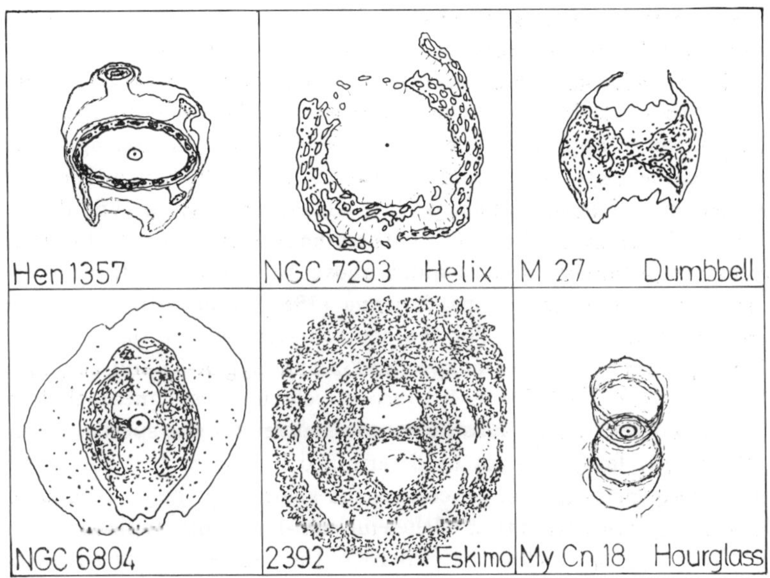

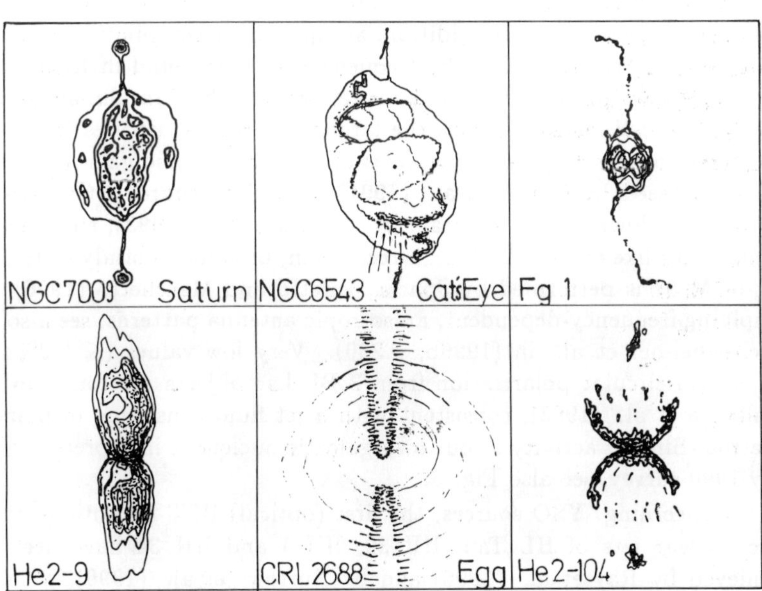

FIGURE 2b. A dozen of suspicious-looking *planetary nebulae*, oriented such that the expected angular-momentum axis is in each case vertical, taken from Kundt (1996b). It is my (often argued) conviction that all of these sources involve pair-plasma jets.

at best to MeV energies. Property (vi) of the jet sources – their occasional *superluminal* appearance – results from a relativistic bulk velocity of the (radiating) jet substance in connection with a small enough inclination of the luminous channel segment to the line of sight. The distribution of apparent superluminal velocities of the extragalactic jet sources leads to *bulk Lorentz factors* above 40 (100 Km/s Mpc H_0): Vermeulen in (Kundt 1996b, p.251), whereby corkscrew-type deviations of the channels from straightness have been ignored. In other words: bulk Lorentz factors much above 10^2 are implied, in agreement with Kundt and Gopal-Krishna (1980), and with Begelman, Rees and Sikora (1994). For the neutron-star sources, on the other hand, apparent velocities of c/4 (SS433 and Cyg X-3) and 2c (GRS 1915+105 and GRO J1655-40) respectively may well mean that in all four sources, the jet substance moves at relativistic speeds. Their bulk Lorentz factor cannot be determined in this way; it should be found from their spectra. Of course, the velocity of the jet substance cannot be guessed from that of the emission-line knots, typically some 10^3 times slower, which is the velocity of the – some 10^9 times heavier – channel-wall material.

Most of what has been said above can be found in greater detail in my 1996b book. In addition, an up-to-date compilation of 55 *quasars* at three different radio frequencies can be found in Reid et al. (1995). 8 Mpc-sized radio galaxies are studied by Subrahmanyan, Saripalli and Hunstead (1996), the largest coherent structures in the Universe. Spectral aging in the lobes of the strong radio galaxy Cyg A is discussed by Carilli et al. (1991). The Mpc-source NGC 315 is resolved down to the pc scale by Venturi et al. (1993), showing inner lobes like in Cen A. A high-resolution, broadband study of the jet of M 87 is performed by Sparks, Biretta and Macchetto (1996), implying frequency-dependent, anisotropic antenna patterns; see also Meisenheimer et al. in (1996b, p.230). Very low values ($\lesssim 0.2\%$) of optical circular polarization from 2 BL Lac objects are found by Valtaoja et al. (1993), consistent with a jet fluid consisting of pair plasma. Similar activity in our own galactic nucleus is interpreted in my 1990 survey, see also Fig. 3.

Concerning YSO sources, the first (optical) HST resolutions of the nuclear jets of HL Tau, HH 30, HH 1 and HH 34 have been achieved by Ray et al. (1996) and by Burrows et al. (1996), with initial jet widths of order 0.1", or $10^{14.3}$cm (!), (10 times smaller than in the first press release). More deeply embedded YSO outflows, like L 1448, have been traced in near-infrared molecular lines: Bally, Lada and Lane (1993). Remarkable excitation of HCO^+ outside the (main) lobes of HH 34 has been found by Rudolph and Welch (1992); is it a Strömgren-type phenomenon? Non-thermal radio emission from near

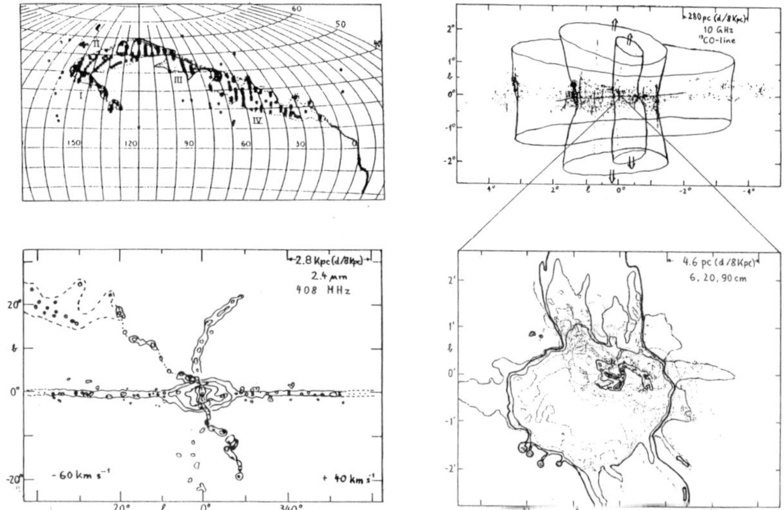

FIGURE 3. Morphology of the twin-jet from Sgr A* in our *Galactic Center*, taken from (Kundt 1996b). Sgr A East apparently serves as the storage bubble in Blandford and Rees's twin-exhaust scenario.

the bow shock of NGC 1977 in Orion B has been detected by Subrahmanyan (yet unpublished).

2. NON-BLACK-HOLE ENGINES AND LARGE BULK LORENTZ FACTORS

How uniform is the jet phenomenon? Can black holes be as efficient as *oblique magnetized rotators* for making jets – like YSOs, white dwarfs and neutron stars? I doubt it. In any case, when Rees (1997) talks about 'Astrophysical Evidence for BHs', he ignores the fact that disc densities grow singular at their origin, i.e. that a compact core is predicted by standard theory at the center of every galactic disc: Kundt (1996a). More in detail, once the surface mass density of a disc approaches stellar values, the disc will enter its main-sequence stage of hydrogen burning – at solar-system distances from its center – and evolve inwards towards episodic nuclear detonations near its center; all this long before the compactness can grow high enough for BH formation. Active galaxies are likely to contain 'burning discs' at their centers, of mass $\gtrsim 10^{6.5} M_\odot$.

And how to *make the jets*? If in-situ electron acceleration rivals with the Second Law (1996b, pp. v, 6, 284), i.e. if the huge Lorentz factors inferred from the spectra must be provided by the central engine, there is a clear indication that the jet substance derives from *coronal* (or magnetospheric) *discharges* whenever their voltage exceeds the pair-formation threshold (of 10^{12}eV(eV/hν)), i.e. whenever the boosted primary charges are energetic enough to create pairs on collision with the embedding photon bath. Centrifugal post-acceleration pressurizes the pairs, and raises their Lorentz factors. That these Lorentz factors do exceed 10^5 or more in the extragalactic sources is already evidenced by the extremely hard spectra of naked-core sources, reaching and exceeding TeV spectral-peak energies. Only some 1% of the initial pair energy escapes through the jets, the remaining 99% are emitted as VHE radiation.

The initial *focusing* of the jets is thus achieved via *ram-pressure confinement* by halo material, according to Blandford and Rees's twin-exhaust mechanism; cf. Fig. 3. The focusing is stable because of the pair plasma's weightlessness and inmiscibility. When friends of mine, like Max Camenzind, claim they can focus stellar winds via their frozen-in magnetic flux, they assume an unrealistically ordered flux. Disaster will set in at the first field reversal – as found in 3D numerical simulations. Only axially symmetric simulations have had success in focusing jets through a medium of comparable composition.

Just as laboratory jets are cooled by shooting at them with a laser beam, the forming cosmic jets are *cooled* by crossing the photon bath of the BLR so that their motion approaches a loss-free $\vec{E} \times \vec{B}$-drift. *Losses* do occur whenever their vacuum channel is bent – or intruded from outside – as well as through inverse-Compton collisions with the residual ambient photon bath. Such unavoidable inverse-Compton losses limit the upper turnover of the electron spectrum, allowing for Lorentz factors as high as 10^7 (even) in the heads of short sources, like Pictor A, but only of 10^5 in those of the Mpc sources (Meisenheimer and Röser 1989). Channel bends, on the other hand, show up as emission knots. The terminal jet of 3C 273 is a lucid case of in-situ deceleration of its ramming plasma: Meisenheimer et al. in (Kundt 1996b, p.230).

If it is true that also stellar jets are rammed by relativistic pair plasma, why don't we see their *lobes* – often called 'cavities' – glow at *radio* frequencies? I think we will, at low enough frequencies. Their spectra indicate Lorentz factors of $\lesssim 10^3$, some 10^3 times lower than those of their extragalactic big brothers, implying 10^6-times softer radiation for similar magnetic-field strengths. As mentioned above, we know already a handful of (expanding) stellar radio triples, and one-sided nuclear jets.

Acknowledgments

My thanks for the hospitable invitation to Torino go to the organizers, and for the manuscript to Harald Giersche and Carsten van de Bruck.

REFERENCES

Bally, J., Lada, E.A., & Lane, A.P., 1993, *Ap.J.*, **418**, 322.

Begelman, M.C., Rees, M.J., & Sikora, M., 1994, *Ap.J.*, **429**, L57.

Blome, H.J., & Kundt, W., 1988, *Ap&SS*, **148**, 343.

Burrows, C.J., Stapelfeldt, K.R., et al., 1996, *Ap.J.*, **473**, 437.

Carilli, C.L., Perley, R.A., Dreher, J.W., & Leahy, J.P., 1991, *Ap.J.*, **383**, 573.

Hirth, G.A., Mundt, R., & Solf, J., 1994, *Ap.J.*, **427**, L99.

Koyama, K., Hamaguchi, K., Ueno, S., Kobayashi, N., & Feigelson, E.D., 1996, *PASJ*, **48**, L87.

Kundt, W., 1990, *Ap&SS*, **172**, 109.

Kundt, W., 1991, *Comments on Astrophysics*, **15**, 255.

Kundt, W., 1996a, *Ap&SS*, **235**, 319.

Kundt, W., 1996b, ed. of: 'Astrophysical Jets and their Engines', Lecture Notes in Physics **471**, Springer.

Kundt, W., 1996c, in: 'Supersoft X-Ray Sources', ed. J. Greiner, Lecture Notes in Physics **472**, Springer, p. 45.

Kundt, W., 1996d, in: 'Disks and Outflows around Young Stars', eds. S. Beckwith, J. Staude, A. Quetz & A. Natta, Lecture Notes in Physics **465**.

Kundt, W., 1997, *Fundamentals of Cosmic Physics*, 1997, submitted.

Kundt, W., Gopal-Krishna, 1980, *Nature*, **288**, 149.

Meisenheimer, K., & Röser, H.-J., 1989, eds. of 'Hot Spots in Extragalactic Radio Sources', Lecture Notes in Physics **327**, Springer.

Ray, T.P., Mund,t R., Dyson, J.E., Falle, S.A.E.G., & Raga, A.C., 1996, *Ap.J.*, **468**, L103.

Rees, M.J., 1997, review lecture at 18th Texas Symposium on Relativistic Astrophysics and Cosmology at Chicago, Dec. 1996, to appear in Ann. N.Y. Acad. Sci.

Reid, A., Shone, D.L., Akujor, C.E., Browne, I.W.A., Murphy, D.W., Pedelty, J., Rudnick, L., & Walsh, D., 1995, *A&AS*, **110**, 213.

Reipurth, B., & Heathcote, S., 1993, in: 'Astrophysical Jets', eds. D. Burgarella, M. Livio, & C.P. O'Dea, Sp. Tel. Sci. Inst., Symp. Ser. **6**, Cambridge, p. 35.

Rodríguez, L.F., & Reipurth, B., 1994, *A&A*, **281**, 882.

Rudolph, A., & Welch, W.J., 1992, *Ap.J.*, **395**, 488.

Sparks, W.B., Biretta, J.A., & Macchetto, F., 1996, *Ap.J.*, **473**, 254.

Subrahmanyan, R., Saripalli, L., & Hunstead, R.W., 1996, *MNRAS*, **279**, 257.

Trammell, S.R., & Goodrich, R.W., 1996, *Ap.J.*, **468**, L107.

Valtaoja, L., Karttunen, H., Valtaoja, E., Shakhovskoy, N.M., & Efimov, Yu.S., 1993, *A&A*, **273**, 393.

Venturi, T., Giovannini, G., Feretti, L., Comoretto, G., & Wehrle, A.E., 1993, *Ap.J.*, **408**, 81.

JETS IN AGN:
THE PARSEC SCALE

CHAPTER 6

GAMMA-RAY OBSERVATIONS OF BLAZARS: WHAT DO THEY TELL US?

G. GHISELLINI

Osservatorio Astronomico di Brera, I-22055 Merate (Lecco), Italy

More than 60 blazars (BL Lacertae objects and flat radio spectrum quasars) have been detected above 100 MeV by the CGRO satellite. The rapid variability implies that the plasma responsible for this emission must have relativistic bulk motion at small angles to the line of sight. A very little fraction of the power emitted in the γ–rays can be reprocessed through γ–$\gamma \to e^+ e^-$ mechanism since the inevitable by–product of this process, i.e. X–rays, should be more than observed. Therefore we argue that the emission region must be located at some hundreds Schwarzschild radii from the putative black hole, and therefore that the transportation of energy up to these distances must be dissipationless. Some recent ideas about the origin of the γ–ray emission are discussed.

1. SPECTRAL ENERGY DISTRIBUTION OF BLAZARS

The overall spectral energy distribution (SED) of the blazars detected by CGRO shows two peaks, in a ν–$\nu F(\nu)$ plot (see e.g. von Montigny et al. 1995). The first peak occurs in the IR–UV band, the second in the MeV–GeV band, with a minimum in the X–ray range (see Fig. 1).

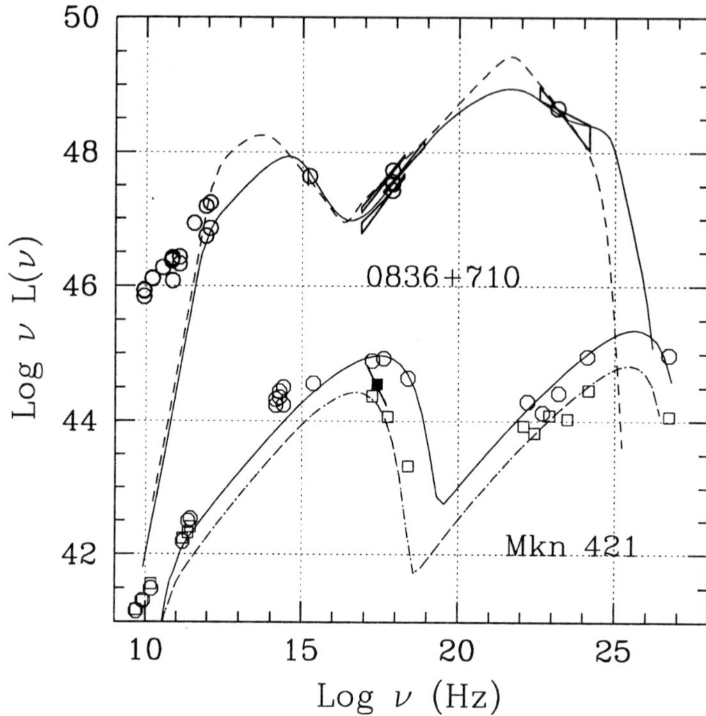

FIGURE 1. This figure shows the overall spectra of two blazars (a BL Lac object and a quasar). Note how the γ-ray emission dominates the power output, especially in the case of 0836+710, the location of the two peaks, and the fact that Mkn 421 has been detected up to TeV energies. Lines correspond to simple one–zone homogeneous models, as explained in Comastri et al. (1997).

Many models have been proposed to account for the blazar SED and the violent behavior of blazars, characterized by factor 2 variability in few hours. All models agree on the non–thermal nature of the underlying plasma, and on the fact that their emission region must move at relativistic speeds ($\beta = v/c \sim 0.995$, corresponding to a bulk Lorentz factor $\Gamma = 10$). Superluminal apparent velocities and the very fact that we observe variable γ-ray emission are dramatic proofs of this assumption. Without bulk motion, and the corresponding beaming of the produced radiation, the large densities of high energy photons would make the source opaque to γ-rays (Dondi and

Ghisellini 1995). In addition, the electron producing the radio emission by the synchrotron process would produce (by the self–Compton process) order of magnitudes more X–rays than observed.

All models agree that the first peak of emission (from radio to EUV) is due to the synchrotron process. They differ on the interpretation of the second peak, where most of the power is emitted, and on the location of the corresponding emission region.

According to one interpretation, the seed photons can be produced by synchrotron emission internally to the emitting blob or jet (Maraschi, Ghisellini and Celotti, 1992). Another possibility is that the blob itself could illuminate a portion of the broad line region, whose reprocessed line radiation can dominate the radiation energy density in the blob (Ghisellini and Madau, 1996). In another scenario the seed photons are produced externally, by the accretion disc (Dermer and Schlickeiser 1993), by the broad line region illuminated by the disc (Sikora, Begelman and Rees 1994) and/or by some scattering material surrounding the jet (Blandford and Levinson 1995). Finally, a dusty torus could provide IR radiation to be scattered in the γ–ray band, as suggested by Wagner et al. (1995), or both peaks could be due to the synchrotron emission of two different (but related) particle distributions, as proposed by Mannheim (1993).

2. LOW ENTROPY INNER JETS

Quite independently on the specific model, a very important conclusion can be reached concerning the location of the active region, which must be at some distance from the central powerhouse. Assume in fact that the site of conversion of primary power into radiation is a compact and inner region of the jet, close to the accretion disc and therefore located in a dense photon environment. In this case part of the γ–ray radiation is absorbed by γ–γ collisions (Blandford and Levinson 1995). The pairs created in this way are relativistic, and emit at lower frequencies. All the absorbed power in the γ–ray band reappears at lower energies, namely in the X–ray band. Therefore it is inevitable to predict that the X–ray luminosity should be of the same order of the γ–ray luminosity. This is not observed.

A way out of this is to assume that the cooling time of the pairs is longer than their escape time. In this case only a fraction of the power absorbed in the γ–ray band is reprocessed into radiation in the X–ray band. This model then requires that there are sufficient X–rays to absorb the γ–rays, but not enough photons for Compton cooling. In principle, this is possible, because the scattering between pairs and

X–rays occurs in the inefficient Klein Nishina regime, but it is highly unlikely, because the X–ray emission should always be accompanied by at least a comparable amount of optical–UV radiation (Ghisellini and Madau 1996). We then conclude that the γ–ray emitting region must be transparent to the $\gamma - \gamma \rightarrow e^{\pm}$ process, and therefore must be located far from the (X–ray emitting) accretion disc. On the other hand, the short variability timescales observed at high energies limit the dimensions of the source, and hence its location in the jet. By combining both limits, we can derive a typical distance at which dissipation occurs, of a few hundreds Schwarzschild radii.

In the inner part of the jet, energy must be transported efficiently, without dissipation. We can put some lower limits on the transported energy, since it must be sufficient to power the extended radio lobes. The energy content of the lobes (calculated i.e. by the minimum energy requirement), divided by their lifetimes yields the average power necessary for the lobes to exist. Note that this power can be much greater than the power radiated in the radio band (Rawlings and Sanders 1991), and can reach values exceeding 10^{47} erg s^{-1}.

Examples of low entropy transportation systems are cold (in the comoving frame) protons with bulk Lorentz factor Γ or Poynting flux.

The kinetic power carried by the protons is

$$P_k = \pi r^2 n_p \Gamma \beta m_p c^3 \rightarrow \tau_p \sim 5 \times 10^{-2} \left(\frac{P_{k,46}}{\Gamma_1 r_{14}} \right)$$

where $r = 10^{14} r_{14}$ cm is the cross sectional radius of the jet, $\tau_p = \sigma_T n_p r$, $\Gamma = 10 \, \Gamma_1$ and $P_k = 10^{46} P_{k,46}$ erg/s. As Sikora et al. (1997) pointed out, cold electrons of the same optical depth would produce an observable bump at ~ 1 keV by scattering ambient UV photons coming from the accretion disc and from the broad line region. This bump is not observed, *excluding cold electrons as a viable possibility*.

Another issue affected by these arguments is the composition of jets. Should they mainly made by electron–positron pairs, then one cannot invoke the reprocessing of a significant part of the γ–ray emission to make them, since we do not see the leftover of the process (i.e. X–rays).

3. WHICH MODEL?

On the basis of simple fits to the overall energy distribution of blazars it is not possible to discriminate among emission models to interpret the entire radiation spectrum. One must therefore find alternative ways. One of these is coordinated variability in different energy bands. In fact different models predict different variability behavior, as in the case of the synchrotron self–Compton model (SSC) and the models in which the seed photons to be Comptonized are produced independently, in an external region of the jet. Assume in fact to vary (increase) the number of emitting electrons. Synchrotron (optical) photons increase linearly with the electron number, while the self–Compton radiation will produce a quadratic variation (there are more scatterings because there are more synchrotron photons *and* more electrons). In the external Compton case, instead, also the Compton radiation will vary linearly with the electron number. Both models would predict (in the simple case of a one–zone emitting region) simultaneous variations, but of different amplitudes. Instead the 'mirror' model, in which the jet itself illuminates a portion of the broad line region, predicts a delay between the optical–UV emission and the γ–ray flux, which lags the optical (Ghisellini and Madau, 1996).

4. A POSSIBLE SEQUENCE

It is conceivable to assume, as a working hypothesis, that for all the lineless BL Lacertae objects the main mechanism at the origin of the high energy emission is the self–Compton process, while in radio–loud quasars the broad emission lines make an important contribution to the seed photons to be Comptonized. Then in the latter class of sources the more abundant seed photons should produce a larger γ–ray luminosity (with respect to the optical). It is therefore gratifying that this is just what we observe. With few exceptions, which perhaps should alert us on the possible misclassification of some of the objects: after all, sources like PKS 0537–441, classified as a BL Lac, sometimes shows prominent emission lines. Going further, we can ask if there are some systematic differences between BL Lacs and quasars. As Comastri et al. (1997) pointed out, there is an anti–correlation between the slopes in the X–ray and the γ–ray bands: sources with a steep α_γ tend to have flatter X–ray slopes, and the BL Lac objects are indeed characterized by the steepest X–ray indices and the flattest α_γ. A deeper analysis, performed by considering the entire spectral energy distribution of all γ–loud blazars, shows that there is a remarkable correlations between the energy of the peak of the synchrotron spectrum and the observed overall luminosity (Ghisellini et al. 1997, in preparation). BL Lacs tend to have their 'synchrotron peak' at energies larger than quasars, and correspondingly also their high energy, 'Compton peak' is often in the GeV or even TeV band. This indicates that the electrons responsible of most of the emission have an energy which is larger in (mostly low luminosity) BL Lacs than in (highly luminous) quasars. It is tempting to speculate that this trend reflects some sort of balance between the cooling suffered by these electrons and the (yet unknown) reacceleration mechanism: when the cooling is larger, the maximum value of the energy attained by the electrons is smaller.

In this scenario, the presence or absence of external seed photons may thus determine the entire blazar spectrum: lineless objects suffer less cooling and their characteristic electron energy is large, making synchrotron radiation to extend to the X–ray band, and the self–Compton emission to the TeV band. Since the Compton cooling is not extreme, the γ–ray luminosity, even if it extends to the TeV band, is not (or only marginally) dominant with respect to the synchrotron luminosity. Only BL Lac objects with the 'synchrotron peak' in the X–rays should therefore be TeV emitters (see Fig. 1).

In objects with broad line emission, Compton cooling is more severe, and the electron energies are smaller. The synchrotron peak lies in the mm–IR part of the spectrum, and the Compton peak in the MeV band. Just because the Compton cooling is larger, these objects are Compton dominated, i.e. their γ to optical luminosity ratio is much greater than unity.

In between we can find intermediate sources, including BL Lac objects showing occasionally broad emission lines.

REFERENCES

Blandford, R.D., 1993, in *Compton Gamma Ray Observatory*, ed. N. Gehrels & M. Friedlander (New York: American Inst. of Physics), p. 533.

Blandford, R.D. & Levinson, A., 1995, *Ap.J.*, **441**, 79.

Comastri A. et al., 1997, *Ap.J.*, **480**, 534.

Dermer, C. & Schlickeiser, R., 1993, *Ap.J.*, **416**, 458.

Dondi, L. & Ghisellini, g., 1995, *MNRAS*, **273**, 583.

Ghisellini, G. & Madau, P., 1996, *MNRAS*, **280**, 67.

Mannheim, K., 1993, *A&A*, **269**, 67.

Maraschi, L., Ghisellini, G. & Celotti, A., 1992, *Ap.J.*, **397**, L5.

Rawlings, S.G. & Saunders, R.D.E., 1991, Nature **349**, 138.

Sikora, M., Begelman, M. & Rees, M.J., 1994, *Ap.J.*, **421**, 153.

Sikora, M. et al., 1997, *Ap.J.*, in press.

von Montigny, C. et al., 1995, *Ap.J.*, **440**, 525.

Wagner, S.J. et al., 1995, *A&A*, **298**, 588.

THE MATTER CONTENT OF JETS IN ACTIVE GALACTIC NUCLEI

A. CELOTTI

S.I.S.S.A., via Beirut 2–4, 34014 Trieste, Italy

The problematic of the matter content of jets in Active Galactic Nuclei is discussed. Through a comparison of models and observations, new pieces of evidence have emerged in the last few years, which give clues on the composition of jets.

1. INTRODUCTION

Among the open problems to which this meeting is dedicated, certainly is the composition of the plasma flowing in jets associated with Active Galaxies (AGN). In 1984, Begelman, Blandford and Rees discussed the major points, mainly based on theoretical arguments, in favor and against the hypothesis of heavy (hereafter referring to electron–proton plasma) and light (electron–positron pair plasma) jets. But no unique conclusion had been reached on their composition.

However, both recent observations with VLBI resolution of a large number of radio sources and the detection of γ–ray emission from several tens of blazars, have allowed us to consider this issue from a more phenomenological point of view and derive some important clues.

In the following, I summarize the main arguments and the conclusions which can be inferred. For simplicity, it is adopted the extreme view that jets are either composed of an electron–positron ($e^{\pm}$) or electron–proton (e–p) plasma.

2. THE SSC MODEL AND VLBI JETS: KINETIC POWER AND PARTICLE FLUX

Most of the simple, order of magnitude, arguments presented here are based on the assumption that parsec scale jet emission is mainly produced by the synchrotron self–Compton (SSC) process. By applying the (standard) model to the observed spectra it is then possible to derive physical quantities, namely the emitting particle density n and the relativistic Doppler factor δ of the jet. These in turn allow one to estimate (on these scales) the power $L_{\rm kin,pc}$ carried along the jet in kinetic form and the particle number flux $\dot{N}_{\rm ssc}$, as

$$L_{\rm kin,pc} \simeq A\, m_e c^3\, \Gamma^2\, \chi\, n\, \gamma_{\rm min}^{-1.5} \qquad \text{and} \qquad \dot{N}_{\rm ssc} \simeq A\, c\, \Gamma\, n\, \gamma_{\rm min}^{-1.5}$$

where A is the jet cross sectional area, estimated from VLBI measures, a typical energy spectral index $\alpha = 0.75$ has been adopted and Γ is the Lorentz factor of the bulk motion (here assumed $\Gamma \sim \delta$). χ parametrises the uncertainty on the plasma composition ($\chi \sim 1$ for an $e^{\pm}$ plasma and $\chi \sim 2 \times 10^3$ for an e–p jet). $\gamma_{\rm min} m_e c^2$ is the minimum energy at which the emitting particle distribution effectively extends as a power–law (Note that this cannot be directly inferred from the observed spectra). n corresponds to the total particle density for $\gamma_{\rm min} \sim 1$.

Indeed, χ and $\gamma_{\rm min}$ result to be the major unknowns in the estimated $L_{\rm kin,pc}$ and $\dot{N}_{\rm ssc}$. *Constraints on $L_{\rm kin,pc}$ and $\dot{N}_{\rm ssc}$ thus turn into constraints on the jet plasma composition and the extension of the emitting particle distribution.*

3. HEAVY JETS IN RADIO–LOUD QUASARS

$L_{\rm kin,pc}$ and $\dot{N}_{\rm ssc}$ can be compared with the analogous quantities on different jet scales. On one side with $L_{\rm kin,kpc}$, i.e. the kinetic power estimated on extended (typically kpc-Mpc) scales, e.g. by Rawlings and Saunders (1991). On the other side with the particle (e–p) flux from the very inner AGN region, as inferred from typical accretion rates, $\dot{N}_{\rm acc} \lesssim 8 \times 10^{49}\, \eta_{0.1}^{-1}\, M_8\, {\rm s}^{-1}$. As reference value, the Eddington rate for a $M = 10^8 M_\odot M_8$ central black hole mass and efficiency $\eta = 0.1\eta_{0.1}$, has been assumed.

The results of such comparisons for a sample of radio loud quasars suggest that (Celotti and Fabian 1993):

1) $\dot{\mathcal{N}}_{\rm ssc} \lesssim \dot{\mathcal{N}}_{\rm acc}$ only for $\gamma_{\rm min}^{1.5} \gtrsim 10^3$

2) $L_{\rm kin,kpc} \sim L_{\rm kin,pc}$, which is indeed what expected, as most of the power on pc scales is likely to be in kinetic form and no large jet power dissipation signature can be found (observed radiative losses can be indeed a minor source of dissipation). This means that $\chi\,\gamma_{\rm min}^{-1.5} \sim 1$ and consequently $\chi \gtrsim 10^3$.

In other words, the number flux of particles which can flow from the central source (presumably as part of the accreting material) to VLBI scales requires that the actual *emitting particles do not cool down to energies below typically $\sim$ 50 MeV (i.e. $\gamma_{\rm min} \sim$ 100). This in turn implies that, in order for jets to transport enough power to large scales with a limited number of particles, they have to be heavy.*

4. THE EMITTING PARTICLE DISTRIBUTION: A $\sim$ 50 MeV CUTOFF ?

Interestingly, as also emerged during this meeting, several independent considerations are in favor of/require $\gamma_{\rm min}$ to be $\gg$ 1 and possibly of the order of $\sim$ 100. These include: Faraday depolarization arguments, estimates of brightness temperatures, arguments derived from MHD models for the formation and propagation of jets (Camenzind, Chapter 1), indications of small amounts of mildly relativistic/thermal particles in jets (Sikora et al. 1997; Celotti et al. 1997), high energy emission (see next Section), estimates of the efficiency of $e^\pm$ annihilation in jets (Ghisellini et al. 1992), observations of emission from hot spots on large scales (Meisenheimer et al. 1989). *(Re-)acceleration/heating/escape processes must efficiently operate.*

5. γ–RAY EMISSION

As presented in Ghisellini's contribution to these proceedings (Chapter 6), the detection of high energy (γ–ray) flux from blazars has revealed new clues on the physics (and composition) of relativistic jets. In particular, for our purposes here, it is worth stressing two points: i) the very same observation of γ–ray photons implies that they are produced by plasma in relativistic motion; ii) the copious production of $e^{\pm}$ pairs in regions dominated by intense radiation fields (such as the Broad Line Region of radio–loud quasars) would give rise to the reprocessing of the emitted spectrum and consequently to an X–ray flux larger than the detected one. *This, coupled with constraints on the amount of thermal/non–relativistic material present in jets, constitutes a further piece of evidence towards relatively low $e^{\pm}$ particle content, at least up to distances 10^2–10^3 times the gravitational radius.*

6. THE CASE OF LOW POWER SOURCES: LIGHT $e^{\pm}$ JETS?

Low luminosity radio–loud sources, namely BL Lac objects and their likely parent population of FR I radio galaxies, present different spectral and morphological properties with respect to the more powerful counterparts of radio–loud quasars and FR II radio galaxies. It is therefore interesting to estimate $L_{\rm kin,pc}$ (and the luminosity in line emission) of low power objects and compare them with those of high power ones.

The main findings of the study of low luminosity sources can be summarized as follows:

1) Low and high power radio–loud AGN seem to have comparable $L_{\rm kin,pc}$ but significantly different total line luminosity (Celotti, Padovani and Ghisellini 1997). We stress that here and in the following $L_{\rm kin,pc}$ is computed by assuming the same value of χ and $\gamma_{\rm min}$ for the two classes.

2) The distributions of $L_{\rm kin,kpc}$ for FR I and FR II radio galaxies, although based on a limited number of objects, are clearly distinct, with the average $L_{\rm kin,kpc}$ differing by about a factor 10^2–10^3 (see Fig. 1).

3) BL Lac objects and radio–loud quasars have indistinguishable (pc scale) kinetic power distributions $L_{\rm kin,pc}$, as shown in Fig. 2.

4) A study of the spectral and energetic properties of the FR I radio galaxy M87 favors the presence of light jets (Reynolds et al. 1996).

Based on these pieces of evidence, we therefore suggest that the jet plasma in low power sources (namely BL Lacs and FR I radio galaxies) is pair dominated. Their kinetic luminosity, as estimated by applying the SSC model, would then be about three orders of magnitude smaller than $L_{\rm kin,pc}$ of powerful AGN, indeed in agreement with the findings on extended scale kinetic powers.

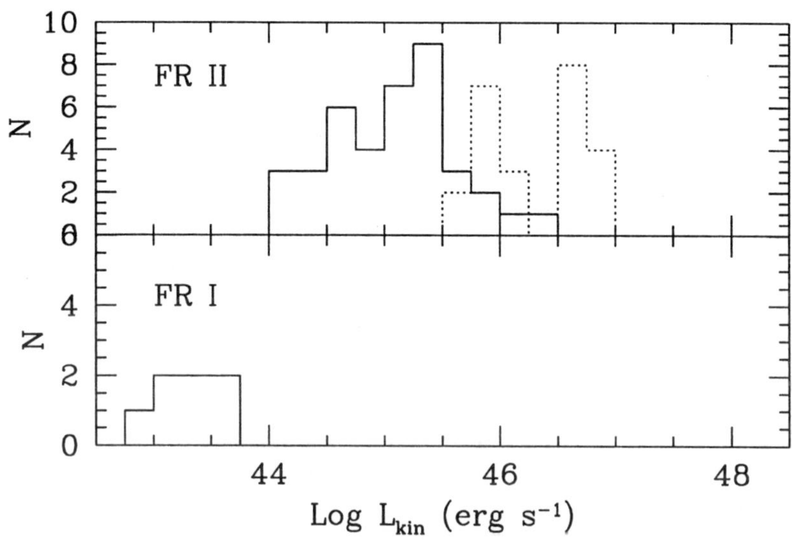

FIGURE 1. The distribution of $L_{kin,kpc}$, as estimated by Rawlings and Saunders (1991) on large scales, for FR II (top panel) and FR I (bottom panel) radio galaxies. Dotted lined represent a non–complete sample.

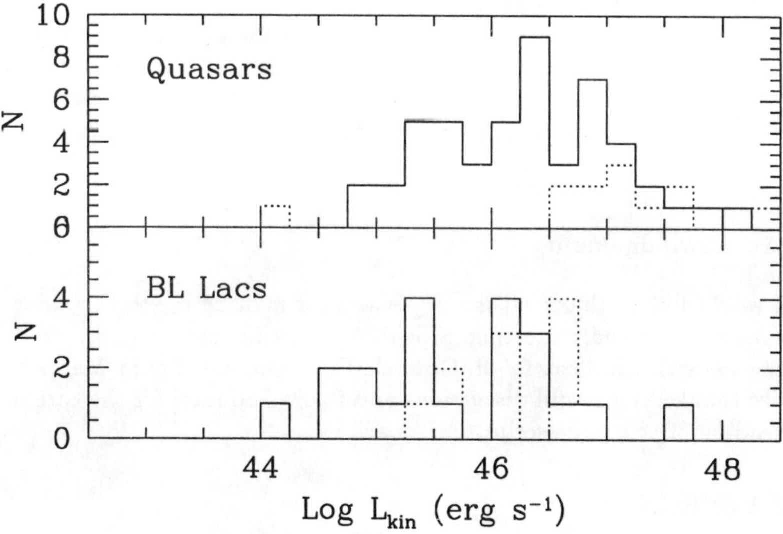

FIGURE 2. The distribution of $L_{kin,pc}$, estimated from the SSC model on pc scale, for radio–loud quasars and BL Lacs (top and bottom panels, respectively). Dotted lines represent objects with $\delta \lesssim 1$, for which L_{kin} has been computed assuming $\Gamma^2 \beta \sim 1$ (see Celotti and Fabian 1993 for details).

It thus appears that *the main difference (most plausibly along a continuous power sequence) in the jet properties of the 'classes' of low and high power radio–loud sources can be in the composition of the jet plasma, with low power objects being dominated by electron–positron pairs* (Bodo et al. 1997, in preparation).

7. SUMMARY

Arguments based on the estimates of the kinetic power and particle number flux on parsec scales suggest that jets of powerful radio sources are mainly composed of an electron–proton plasma. These estimates are mostly derived from the adoption of the SSC model to interpret VLBI and high energy observations (analogous assessments based upon other emission models will be presented in Ghisellini et al., in preparation).

On the contrary, there is some evidence that low power radio sources, namely BL Lac objects and FR I radio galaxies, might be mainly comprised of an electron–positron plasma. It is possible that the plasma composition is the origin of the morphological differences between FR I and FR II jets.

Finally, several arguments set limits on the amount of cold and low energy non–thermal material present in relativistic jets and suggest that the (observed) non–thermal particle distribution does not extend below energies of few tens of MeV.

Acknowledgements

I would like to thank all the organizers for inviting me to this interesting and friendly meeting, providing financial support and... feeding us with excellent food! Gabriele Ghisellini and Laura Maraschi are thanked for useful discussions and Paolo Padovani for the careful reading of this manuscript.

REFERENCES

Begelman, M.C., Blandford, R.D., & Rees, M.J., 1984, *Rev. Mod. Phys.*, **56**, 255.
Celotti, A. & Fabian, A.C., 1993, *MNRAS*, **264**, 228.
Celotti, A., Padovani, P., & Ghisellini, G., 1997, *MNRAS*, **286**, 415.
Celotti, A., Kuncic, Z., Rees, M.J.,& Wardle, J.F.C., 1997, *MNRAS*,

submitted.

Ghisellini, G., & Madau, P., 1996, *MNRAS*, **280**, 67.

Meisenheimer, K., et al. 1989, *A&A*, **219**, 63.

Rawlings, S.G. & Saunders, R.D.E., 1991, *Nature*, **349**, 138.

Reynolds, C.S., Fabian, A.C., Celotti, A., & Rees, M.J., 1996, *MN-RAS*, **283**, 873.

Sikora, M., Madejski, G., Moderski, R., & Poutanen, J., 1997, *ApJ*, in press.

CHAPTER 8

LABORATORY EXPERIMENTS AFFECTING THEORIES OF INTRADAY VARIABILITY IN ACTIVE GALACTIC NUCLEI

G. BENFORD[1] AND H. LESCH[2]

[1]*Department of Physics and Astronomy, University of California, Irvine, USA*

[2]*Institut für Astronomie und Astrophysik der Universität München, Scheinerstraße 1, 81679 München, Germany*

The high brightness temperatures (T_b) implied by quasar intraday variability may be explained by coherent emission, or else by physically implausible bulk relativistic Lorentz factors $\Gamma \geq 100$. Previous theory asserts that various absorption mechanisms will block escape of such coherent, high brightness sources. Yet this same theory fails to account, by many orders of magnitude, for laboratory experiments detecting collective emission. Probably this is because present theory is inadequate.

1. INTRODUCTION

Intraday variability (IDV) in active galactic nuclei can arise from collective emission. (see Wagner and Witzel 1996). Incoherent models demand Doppler factors ~ 100; we have a gap of two to three orders of magnitude between the allowed T_b of incoherently radiating sources and the observations (Romero et al. 1995, Begelman et al. 1994).

A completely different point of view embraces coherent emission (Baker et al. 1988, Krishnan and Wiita 1990, Benford 1984 and 1992b, Lesch and Pohl, 1992). Coherent radiation easily accounts for huge T_b, since the radiation intensity is due to collective emission of all particles within a coherent volume. The rather involved physics is well known from plasma laboratory experiments. The central argument against coherent radiation processes in AGN is saturation of the brightness temperature by induced Compton scattering and/or Raman scattering (Coppi et al. 1993; Levinson and Blandford 1995).

Yet coherent emission appears in the laboratory under conditions which recent astrophysical theory says should not display such radiation. We address this last contradiction, arguing that existing theory of plasma absorption of high brightness temperature emission is inadequate. Since we cannot hope to observe the underlying turbulence in astrophysical cases, laboratory work remains a guide constraining imagination and theory.

The experiments discovering (Kato et al. 1983) and confirming (DiCapua et al. 1988; Yoshikawa et al. 1993 and 1994) strong collective emission at Compton-boosted frequencies share a general approach. Thin relativistic beams of 1-2 MeV ($\gamma = 3$ to 5) propagate through surrounding plasmas whose density exceeds the beam density by several orders of magnitude, $n_b/n_p \sim 0.1$ to 0.001. For higher n_b/n_p strong electromagnetic emission of broadband spectra ~ 1 to 100 GHz appear within a few nsec. of beam introduction, in a narrow cone of angle $\sim 1/\gamma$. Powers exceed a MW.

A later theoretical model for the underlying processes (Weatherall and Benford 1991) invoked collective scattering of beam electrons from strongly concentrated Langmuir turbulence, such as the 'cavitons' often seen in strong turbulence experiments (Goldman 1984; Newman 1985). They proposed a boundary curve within which high beam γ and high n_b/n_p allowed intense emission. This curve (Figure 1, Benford 1992a) can be written in terms of a minimum beam relativistic factor γ_{crit} necessary for collective high-frequency emission at a given beam density ratio,

$$\gamma^2_{crit} = n_p/n_b \ . \tag{1}$$

Thus small beam densities can still induce high frequency emission if sufficiently relativistic. Implications for quasar jets are considerable: quite low beam densities can produce coherent emission if the local (not overall) γ is very high, and thus particularly near acceleration sites. Of course, this has been checked only in experiments at low $\gamma < 4$.

Sometimes dubbed 'superstrong turbulence', in this regime the electrostatic energy density can exceed the local plasma thermal energy density, $W = \langle E^2 \rangle / 4\pi nkT > 1$.

Levron et al. (1988) found that such intense sites occupied a fraction $f \sim 0.01$ to 0.1 of the plasma volume. Since later experiments (Di Capua et al. 1988; Yoshikawa et al. 1993 and 1994) found essentially the same general features as Kato et al. (1983), we shall concentrate on the well-reported Kato et al. results.

The brightness temperature of this emission can be naively estimated from the customary black body formula, with frequency ν_9 in GHz and system size L (all subscripts indicate cgs units)

$$T_b \sim \frac{10^{23} \text{ K}}{\nu_9^2 L_2} \left(\frac{P}{100MW} \right) . \tag{2}$$

A more detailed treatment (Benford and Smith 1982) for emission from a bath of electrostatic waves yields

$$T_b \approx 10^{27} \text{ K} \left(\frac{P}{100MW} \right) \frac{(\gamma/3)^2}{n_{12} T_{eV} A_1 (\Delta\nu/10^{11})} . \tag{3}$$

In these formulae we have scaled to the Kato et al. experiments. Here plasma density (n) temperature (T), emitting area (A), frequency (ν) and observed spectral range $(\Delta\nu)$ are in cgs units, with measured power in MW.

The usual Thomson scattering length is $\lambda_T = 1.5 \times 10^{14} n_{10}^{-1}$ cm, which yields for experiment $\sim 10^{10}$ cm and for the quasar case a density limit $n_{10} < R_{15}$, an ample density for radii $R_{15} \sim 1$. More critical limits emerge from inverse Compton and Raman effects.

Levinson and Blandford (1995) studied stimulated Raman scattering (SRS) as a limit to coherent emission, employing a weak turbulence approach. Neglecting magnetic fields, they found that high brightness temperatures T_b inhibited propagation of coherent emission.

Their condition for free-free absorption to dominate either weak stimulated Raman scattering or induced Compton scattering was

$$\frac{n_6}{T_{b12}\nu_9^2 T_6^{3/2}} > 0.16 . \tag{4}$$

There was no Landau damping of the induced Langmuir waves if

$$\frac{n_6}{\nu_9^2} > 60 \, T_6 . \tag{5}$$

For stimulated Raman scattering to dominate demands

$$\frac{n_6 T_{b12}}{\nu_9^2} > 2 \times 10^5 \,, \tag{6}$$

while the single-particle condition for Thomson scattering to be significant is $T_{b12} > 50$. For free-free absorption of the low frequency spectrum,

$$4 \times 10^{-3} \frac{\nu_9^2}{n_6} T_6^{3/2} < 1 \,. \tag{7}$$

In the experiments of Kato et al. condition (4) fails, whereas (5) is valid. Free-free absorption should be unimportant and strong Langmuir turbulence can appear, both as observed. Condition (6) works, if $T_b < 10^{31} K$, so stimulated Raman scattering dominates and Thomson scattering can produce strong spectral distortion, since observed $T_b \gg 10^{14}$. Further, the Levinson-Blandford relation for weak Raman back-scattering yields a reflection length

$$L_R \approx 2 \times 10^{-9} \mathrm{cm} \frac{n_{13}(T/eV)^{-3/2}}{\nu_9(T_b/10^{20}\mathrm{K})^2} \,, \tag{8}$$

i.e., ten orders of magnitude less than the length traversed by the observed emission, ~ 1 meter. Since $T_{eV} \sim 1$ to 10 and other parameters are of order unity, weak Raman scattering is plainly ruled out by experiment.

This displays how coefficients based on weak scattering theory can dramatically differ from the reality of strongly coherent laboratory systems. In the experiments collective effects produced strong electromagnetic emission in the 1 to 100 GHz bands, which then propagated through cool surrounding plasma, finally passing through transparent chamber walls to be received by a variety of directional horn antennas. This echoes the general conditions expected near jets leaving quasars, with a plasma screen within which Raman instability is possible.

What is wrong here? Growth rates for Raman instability decline as the emission broadens, and weak turbulence theory assumes that field amplitudes are small. Strong field, broad- band emission violates such assumptions.

Extending such theory to the extreme brightness temperatures and band characteristics of strongly turbulent sources is fraught with peril. Comparison with experiment shows that such results can be misleading.

Anisotropic induced Compton scattering can also occur near quasar jets (Wilson 1982). Coppi, Blandford and Rees (CBR, 1993) considered degrading of a primary emitted photon beam which self-induces Compton scattering. Their approach relied upon weak turbulence theory in which wave-particle interactions dominate wave-wave energy exchanges, as is appropriate for small amplitudes. They roughly estimate the maximum T_b which can escape from a screening plasma as

$$T_b \leq \left[3 \left(\frac{\Omega}{4\pi} \right)^2 n \sigma_T L \right]^{-1} \sim \left(\frac{\gamma}{3} \right)^2 \frac{2.5 \times 10^{19}}{L_2 n_{13}} \text{ K} , \qquad (9)$$

where the Kato et al. experiments observe through a typical propagation length in cool plasma of $L = L_2/100$ cm. Plainly this fails, as experiments routinely observed $T_b \gg 10^{20}$ K.

Perhaps the difficulty with the CBR approach lies in the neglect of collective effects among the scattering electrons. If they do not respond independently to the high brightness flux, the customary scattering coefficients do not apply. As well, CBR ignore magnetic field effects. This is probably risky for jets, which often seem magnetically confined. However, the laboratory experiments intentionally use rather weak magnetic fields, yet still display very high T_b.

Converting their constraint to a density argument, CBR set a limit on plasma density

$$n_p < \frac{2 \text{ cm}^{-3}}{(\Omega/4\pi)^2 T_{b18} R_{15}} , \qquad (10)$$

where the solid angle for emission in experiment is $\Omega \sim 1/\gamma$ and $T_{b18} = (T_b/10^{18}$ K$)$. (Eq. 16 of CBR contains an error in the exponent of R_{15}.) Since emission emerged from a plasma of density $\sim 10^{13}$ cm^{-3}, and column length $R \sim 100$ cm, this estimate is significant. For the laboratory we estimate $R_{15} = 10^{-13}, \Omega = 1/\gamma = 1/3, T_b \sim 10^{27}$, from Eq. (3). Then Eq. (10) yields a plain contradiction,

$$n_p < \frac{3 \times 10^7 cm^{-3}}{T_{b27} L_2} , \qquad (11)$$

Even if we take the customary estimate of Eq. (2), $T_b \sim 10^{23}$, then $n_p < 3 \times 10^{11}$, still short of the $n_p \cong 10^{13}$ through which emission emerged in a clear $1/\gamma$ cone centered on the average pitch angle of the relativistic beam electrons.

2. DISCUSSION

Confrontation of induced Compton absorption and other estimates with plasma experiment suggests that there is no saturation effect for high T_b. Plainly saturation does not limit T_B^{FEL} the brightness of free electron laser-like devices, which often exceeds $10^{30}K$ in plasma environments. Further, radiation is very anisotropic $(1/\gamma \sim \theta)$, so efficiency of induced Compton scattering is almost zero if $\gamma \gg 1$ (Zeldovich et al. 1972). However, this alone cannot explain contradictions with present experiments with γ of 3 to 5.

The thrust of these weak turbulence approaches has been to find screening interactions which prevent escape of high brightness emission. At the root of their difficulties surely lie several approximations:

(a) non adequate treatment of anisotropy (which in quasars is extreme);

(b) simple treatment of the microscopic plasma response to strong waves;

and, especially,

(c) neglect of the macroscopic response of background plasma to the energy and momentum of high brightness radiation, if it is absorbed.

The primary difficulty of these theoretical approaches lies in

(a) their inability to describe existing experiments in coherent emission, due to

(b) their reliance upon weak turbulence theory and isotropic approximations as a guide to strongly turbulent processes, leading to

(c) no explanation for the observed intra-day variability aside from the implausible invoking of bulk relativistic factors $\Gamma > 100$.

These seem grounds to expect that a fully strongly turbulent analysis of the many processes available in such quasar environs will clarify the means whereby coherent emission may escape. Reliance upon experiment can tell us much about the genuinely nonlinear effects expected in powerful, relativistic astrophysical plasmas.

REFERENCES

Baker, D.N., Borovsky, J.E., Benford, G., & Eilek, J.A., 1988, *Ap.J.* **326**, 110.

Begelman, M.C., Rees, M.J., & Sikora, M., 1994, *Ap.J.*, **429**, L57.

Benford, G., & Smith, D.F., 1982, *Phys. Fluids*, **25**, 1450.

Benford, G., 1984, in *Physics of Energy Transport in AGNs*, NRAO Proceedings , p.284.

Benford, G., 1992a, *IEEE Plasma Science*, **20**, 3.

Benford, G., 1992b, *Ap.J.*, **391**, L59.

Coppi, P., Blandford, R.D., & Rees, M.J., 1993, *MNRAS*, **262**, 603.

DiCapua, M.S., Camacho, E.S., Fulkerson, E.S., & Meeker, D., 1988, *IEEE Trans. Plasma Sci.*, **16**, 217.

Goldman, M.V., 1984, *Rev. Mod. Phys.*, **56**, 709.

Kato, K.G., Benford, G., & Tzach, D., 1983, *Phys. Fluids*, **26**, 3636.

Lesch, H., & Pohl, M., 1992, *A&A*, **254**, 19.

Levinson, A., & Blandford, R.D., 1995, *MNRAS*, **274**, 717.

Levron, D., Benford, G., Baranga, A., & Means, J., 1988, *Phys. Fluids*, **31**, 2026.

Newman, D.L., 1985, *Phys. Fluids*, **28**, 1482.

Romero, G.E., Surpi, G., & Vucetich, H., 1995, *A&A*, **301**, 641.

Smith, D., & Benford, G., 1982, *Phys. Fluids*, **25**, 1450.

Wagner, S.J., & Witzel, A., 1996, *A&A*, **36**, 475.

Weatherall, J., & Benford, G., 1991, *Ap.J.*, **378**, 543.

Wilson, D.B., 1982, *MNRAS*, **200**, 881.

Yoshikawa, M., Ando, R., & Masuzaki, M., 1993, *Jpn. J. Appl. Phys.*, **32**, 969.

Yoshikawa, M., Masuzaki, M., & Ando, R., 1994, *J. Phys. Soc. Jpn.*, **64**, 3303.

Zeldovich, Ia.B., Levich, E.V., & Syunyaev, R.A., 1972, *Sov. Phys. JETP*, **35**, No. 4.

HYDRODYNAMICAL MODELS OF SUPERLUMINAL SOURCES

J. L. GOMEZ[1], J. Mª MARTI[2],
Λ. P. MARSCHER[3], J. Mª IBAÑEZ[2],
A. ALBERDI[1]

[1]*Instituto de Astrofísica de Andalucía, CSIC, Apdo. 3004, Granada 18080, Spain*

[2]*Departamento de Astronomía y Astrofísica, Universidad de Valencia, Burjassot 46100, Spain*

[3]*Department of Astronomy, Boston University, 725 Commonwealth Avenue, Boston, MA 02215, USA*

We present numerical simulations of the generation, evolution and radio emission of superluminal components in relativistic jets. The fluid dynamics are computed using a 2D relativistic time–dependent code based on a high resolution shock–capturing scheme, and then used to calculate the radio emission by integrating the transfer equations for synchrotron radiation. These simulations show that a temporary increase in the flow velocity at the jet inlet results in a moving shock which appears in the simulated radio maps as a region of enhanced emission moving downstream at a superluminal apparent velocity. Interactions of the shock with the underlying steady jet result in changes in the internal brightness distribution of the superluminal component, which are manifested as low-level fluctuations about the long–term evolution of both the apparent velocity and the exponential decay of the light curves.

1. INTRODUCTION

The ejection of radiating plasma at apparent superluminal speeds is commonly observed in the radio jets of many blazars, and was recently discovered in galactic sources (e.g., GRO J655–40 [Tingay et al. 1995]). Apparent transverse motions at speeds larger than light can be explained by relativistic fluid motions in jets viewed at small angles to the line of sight. These ejections are usually preceded by outbursts in emission at radio wavelengths, whose frequency–dependent light curves of both total and polarized flux are successfully interpreted in terms of traveling shock waves, based on simplified hydrodynamical models (Marscher and Gear 1985; Hughes, Aller and Aller 1991; Gómez et al. 1994). The recent development of time–dependent hydrodynamical relativistic codes (Martí, Müller and Ibáñez 1994; Duncan and Hughes 1994) makes possible a more detailed description of the generation, structure, evolution, and influence on the radio emission of these shocks (Gómez et al. 1996, 1997; Mioduszewski, Hughes and Duncan 1997; Komissarov and Falle 1996). In this work, we study the ejection, structure, and evolution of superluminal components through variations in the plasma ejection velocity.

2. HYDRODYNAMICAL MODELS OF SUPERLUMINAL SOURCES

2.1. Steady Jet Models

Our fluid dynamics calculations relies on a relativistic, axially–symmetric jet model obtained by means of a high–resolution shock capturing scheme (Martí et al. 1995, 1997) to solve the equations of relativistic hydrodynamics in cylindrical coordinates. In our model, the jet material is represented by an ideal gas of adiabatic exponent $4/3$ and the quiescent state corresponds to a diffuse ($\rho_b/\rho_a = 10^{-3}$), relativistic ($\Gamma_b = 4$), cylindrical beam with (local) Mach number $M_b = 1.69$. (Here ρ stands for proper rest–mass density and Γ for Lorentz factor; subscripts a and b refer, respectively, to atmosphere and beam; values correspond to the injection position). The jet propagates through a pressure–decreasing atmosphere following the law

$$p(z) = \frac{p_a}{[1 + (z/z_c)^n]^{(m/n)}},$$

where $z_c = 60 R_b$ (R_b is the beam cross–sectional radius at injection position) represents the pressure 'scale height' in the axial direction,

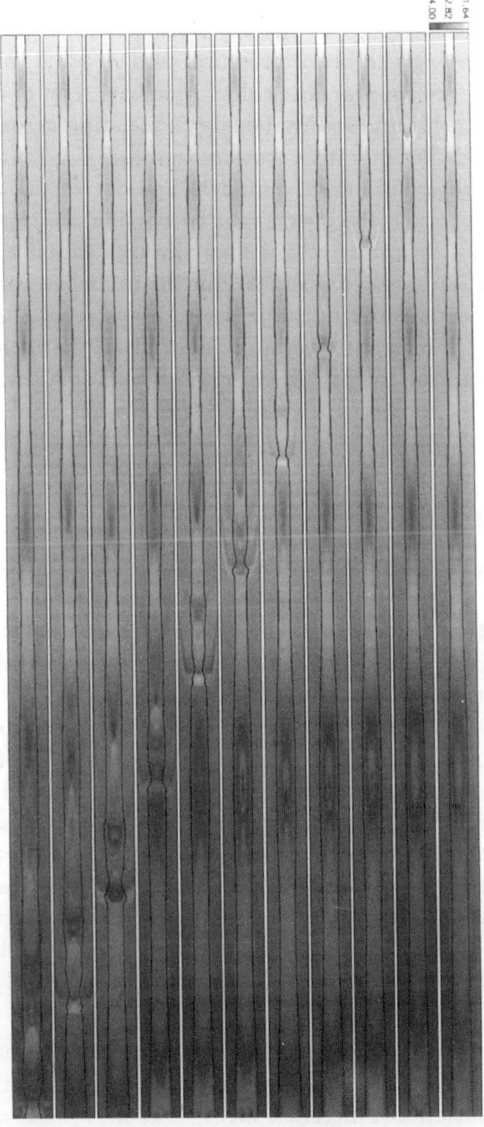

FIGURE 1. Pressure distribution at eleven epochs (0 to 200 R_b/c in steps of 20) after the introduction of a square-wave perturbation to the flow Lorentz factor for the jet model discussed in the text. Owing to the decreasing pressure in the atmosphere, the jet expands with a mean opening half-angle of $\sim 0.4^\circ$. The simulation has been performed over a grid of 1600×80 cells, with a spatial resolution of 8 cells/R_b in both radial and axial directions. Jet boundary is shown with a black curve (see Color Plate 1).

p_a is the ambient pressure at the injection position (in our model, $p_b = 3\,p_a/2$, p_b being the beam pressure) and $n = 1.5$, $m = 2.3$. To obtain the steady jet, we start the calculation with a jet of constant radius, velocity, density and internal energy extending across the whole grid, then the code is run until a stationary solution is found. The top panel in Fig. 1 shows the steady model. The decreasing atmosphere causes the beam to become overpressured outside the injection point and pressure equilibrium is established through radial expansion of the jet. The initial pressure mismatch in the model causes recollimation shocks and expansions in the jet flow. The strength, spacing and periodicity of this internal structure (as well as the local jet opening angle) depend on the jet Mach number and the gradient in the external pressure (Daly and Marscher 1988; Gómez et al. 1995).

2.2. Generation and Evolution of Traveling Perturbations

Once the steady jet model has been obtained, traveling perturbations can be introduced by varying the jet input conditions. We concentrate our attention on the study of the evolution of the flow after the introduction of a square–wave increase of the beam flow velocity from the quiescent value $\Gamma_b = 4$, to $\Gamma_p = 10$ during a short period of time $\tau_p = 0.75 R_b/c$. Because of the faster flow velocity in the perturbation, the fluid in front piles up, creating a shocked state (S) in which the pressure and the rest–mass density are higher than in the quiescent state. The leading shock propagates at $\Gamma_S = 10$. A rarefied state (R) develops behind the shock. The pressure mismatch between these components (S,R) and the external medium causes the transverse section of the beam to change with respect to the steady model. The shocked region expands laterally into the external medium, which has a pressure p_a smaller than p_S. In the rarefied region, on the contrary the beam reduces its radius, since $p_R < p_a$.

The resulting dynamical evolution of the perturbation along the jet is shown in Fig. 1, which contains a set of panels showing the pressure distribution at different epochs. The first panel corresponds to the quiescent jet. Both the shocked and rarefied regions in the perturbation are clearly seen. When the perturbation passes through a standing shock, the latter is 'dragged' downstream for some distance before returning to its initial position as the steady jet becomes reestablished. The interaction with a rarefaction is similar, with the additional feature that a reverse shock is formed, propagating relatively slowly down the jet, before coming to a stop and dissipating (see also Gómez et al. 1996, 1997). The variation in beam radius due to the passage of the component is also evident. This perturbation occurring on the beam surface saturates in small oblique shocks within the beam and is responsible for small variations trailing the main perturbation near the beam axis. The passage of a more powerful component could excite beam radial modes at amplitudes high enough to produce stronger internal shocks.

Within this model resolution, the component pattern speed is compatible with a Lorentz factor of 10, and there are also some indications of quasi–periodic variations, associated with the passage of the perturbation through faster (more rarefied) and slower (more compressed) regions of the previously undisturbed flow. These variations in the pattern Lorentz factor may provide an alternative to geometric effects in the interpretation of the variations in the apparent motion found in many jets (e.g., 3C 345, 0836+71, 3C 454.3, 3C 273, 4C 39.25).

The evolution of the perturbation seems to proceed without noticeable variations in the number of particles contained within it, and maintains approximate self-similarity. It is also consistent with an adiabatic evolution, in concordance with the assumption of previous works (Marscher and Gear 1985; Hughes, Aller and Aller 1989; Gómez et al. 1994).

3. RADIO EMISSION

To compute the radio emission from the jet whose hydrodynamics is calculated as above we distribute the internal energy among the relativistic electrons following the usual power law $N(E)dE = N_o E^{-p}dE$, with $E_{min} \leq E \leq E_{max}$, and spectral index p. Neglecting radiative energy losses, the ratio C_E between the maximum and minimum energy remains constant, and the power law is fully determined by the equations (see Gómez et al. 1995 for more details)

$$N_o = \left[\frac{U\,(p-2)}{1-C_E^{2-p}}\right]^{p-1}\left[\frac{1-C_E^{1-p}}{N\,(p-1)}\right]^{p-2}$$

and

$$E_{min} = \frac{U}{N}\frac{p-2}{p-1}\frac{1-C_E^{1-p}}{1-C_E^{2-p}}$$

where U and N represent the electron energy density and number density, respectively, and are calculated by the hydrodynamical code. Following Wilson and Scheuer (1983), we assume that the magnetic energy density remains a fixed fraction of the particle energy density, which leads to a field with magnitude proportional to $\sqrt{U}$. To account for the small degree of linear polarization observed in many sources, we assume that the magnetic field is predominantly turbulent, but with a weak component parallel to the jet flow.

The emission and absorption coefficients, respectively, for the synchrotron radiation are then computed *in the fluid frame* using (see also Pacholczyk 1970 and Gómez et al. 1993)

$$\varepsilon_\nu^{(\pm)} = \frac{\sqrt{3}}{16\pi}\frac{e^3}{mc^2}\, C_1^{(p-1)/2}\, N_o\, (B\sin\vartheta)^{(p+1)/2}\, \nu^{(1-p)/2}.$$

$$\int_{x_{min}}^{x_{max}} x^{(p-3)/2}\left[F(x)\pm G(x)\right]\mathrm{d}x$$

$$\kappa_\nu^{(\pm)} = \frac{\sqrt{3}e^3}{16\pi m}\,(p+2)\,C_1^{p/2}\,N_o\,(B\sin\vartheta)^{(p+2)/2}\,\nu^{-(p+4)/2}.$$

$$\int_{x_{min}}^{x_{max}} x^{(p-2)/2}\left[F(x)\pm G(x)\right]\mathrm{d}x$$

where ϑ is the angle between the magnetic field and the line of sight; the plus and minus signs are to be taken for the different polarizations, and

$$C_1 = \frac{3e}{4\pi m^3 c^5}$$

$$x = \frac{\nu}{C_1 B\sin\vartheta E^2}$$

$$F(x) = x\int_x^\infty K_{5/3}(z)dz$$

$$G(x) = xK_{2/3}(x)$$

where $K_{5/3}$ and $K_{2/3}$ are the corresponding Bessel functions. These coefficients are then transformed into the observer's frame using the standard Lorentz transformations

$$\varepsilon^{(ob)}\!\left(\nu^{(ob)}\right) = \delta^2\varepsilon(\nu)$$

$$\kappa^{(ob)}\!\left(\nu^{(ob)}\right) = \delta^{-1}\kappa(\nu)$$

where the Doppler factor is $\delta = \Gamma^{-1}(1-\beta\cos\theta)^{-1} = \nu^{(ob)}/\nu$; β is the speed of the fluid in units of c; and θ is the angle between β and the line of sight. Note that both θ and ϑ are corrected by the aberration of light (see Gómez et al. 1994). Also note that all the parameters are cell and *time dependent*, therefore to account for the delays within the jet they are calculated using a retarded time

$$\tau = t - \frac{\vec{x}\cdot\vec{l}}{c}$$

where $\vec{x}$ is the position vector of the cell and $\vec{l}$ denotes the line of sight unity vector.

Once the coefficients are expressed in the observer's frame we solve the transfer equations for the Stokes parameters that determine the radiation field along integration columns parallel to the line of sight following Gómez et al. (1993).

3.1. Steady Jet Emission

Figure 2 shows the total intensity maps corresponding to the stationary model (left panel), and four epochs in the evolution of the disturbance along the jet, with the jet observed at a viewing angle $\theta = 10°$ and optically thin frequency of observation of 43 GHz. To compute the emission we used the following values: $p = 2.2$; $B = 20$ mGauss; external density of $\rho_a = 2 \times 10^{-21}$ g/cm^3; jet radius of 0.1 pc; red-shift of 0.5; Hubble constant of 80 km s^{-1} Mpc^{-1}; and $C_E = 10^3$. Superimposed are contours showing the emission convolved with a circular Gaussian beam as would occur in actual VLBI observations. By looking at the unconvolved stationary map, we observe a regular pattern of knots of high emission, associated with the increased specific internal energy and rest-mass density of internal oblique shocks produced by the initial overpressure in this model. The intensity of the knots decreases along the jet due to the expansion resulting from the gradient in external pressure. VLBI cores can be interpreted as the first of the recollimation shocks in the steady jet (Daly and Marscher 1988; Gómez et al. 1995, 1997). As is illustrated by the convolved maps, the observation of these stationary structures would require very high linear resolution, now achievable through millimeter-wave and space VLBI observations. The first indications of these stationary components have been found in the radio jet of M87 (Junor and Biretta 1995) – as expected due to its relative proximity – and the superluminal galactic source GRO J1655-40 (Tingay et al. 1995). Stationary components have been found in several superluminal sources (e.g., 4C 39.25 (Alberdi et al. 1993), 0735+178 (Gabuzda et al. 1994), 3C 279 (Wehrle et al. 1997)), which may be associated not only with periodic recollimation shocks, but also with bends in the jet (Gómez, Alberdi and Marcaide 1993), or sudden variations in the external pressure that result in strong, isolated recollimation shocks.

3.2. Superluminal Radio Components

In the convolved maps in Fig. 2, we recognize the usual core–jet VLBI

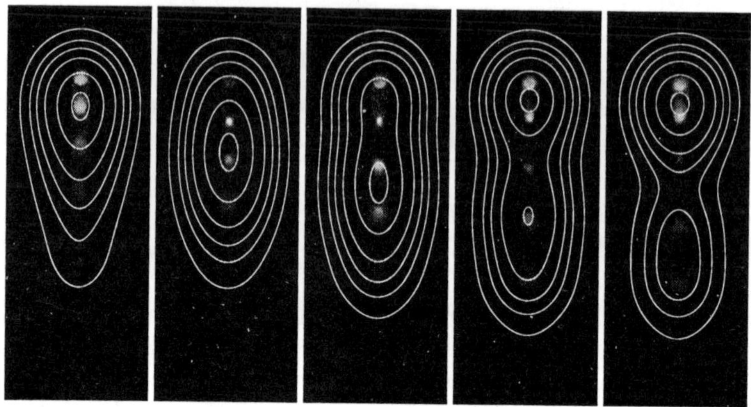

FIGURE 2. Total intensity radio maps at epochs 0.0, 0.9, 1.1, 1.3, and 1.5 years (from left to right) in the observer's rest frame after the introduction of the perturbation. Maps correspond to optically thin 43 GHz observing frequency, and $10°$ viewing angle. Contours show the emission convolved with a circular Gaussian beam of FWHM 0.15 mas. Contour levels are 10, 20, 30, 45, 70, and 95% of the peak intensity. The jet inlet is at the top of each panel. The maps extend 0.9 mas vertically.

structure of a blazar, with a single well-defined, superluminally moving component associated with the moving shock. Note that because the maps are shown for an optically thin frequency of observation, the core reflects the ad hoc jet inlet. The unconvolved maps show a much more complex jet structure. Due to the time delays, the shocked region appears as a very extended region of higher emission (Gómez et al. 1994, 1997), which is moving and interacting with the quiescent jet. As a result of this interaction, the previously stationary hot-spots in the jet increase in flux, being later dragged temporarily downstream. Dragging in component K1 of 0735+178 has been observed (Gabuzda et al. 1994), and new VLBI observations of this source and the radio galaxy 3C 120 are underway to compare this behavior more closely with our model. Similar results of the interaction of superluminal and stationary components have also been observed in 3C 279 (Wehrle et al. 1997), and may be expected in other sources as more high–frequency images become available.

Figure 3a shows the separation of the brightness centroid of the moving component as a function of time. Fitting the separation of the brightness centroid versus time to a straight line starting at epoch 1.0 yrs, when the new component first becomes distinguishable from the core in the maps, we obtain an apparent velocity of ~ 8.6 c, in general concordance with the expected value for a pattern Lorentz

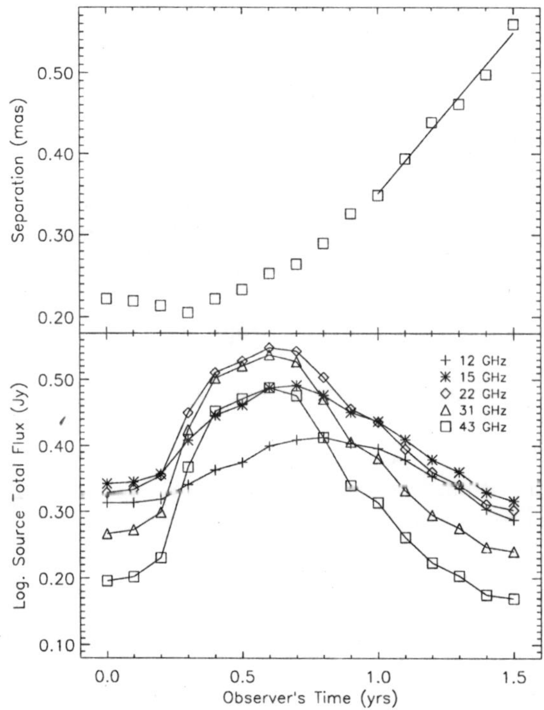

FIGURE 3. (*a, top*) Separation of the component observed in the convolved maps of Fig. 2 vs. time. Fitting epochs 1.0 to 1.5 yrs to a straight line gives an apparent velocity of 8.6 c. (*b, bottom*) Simulated multifrequency light curves for the model.

factor of $\Gamma \sim 10$ and viewing angle of $10°$.

Light curves covering the evolution are shown in Fig. 3b for five different frequencies. The ejection of the component is accompanied by a burst in flux at all frequencies. The subsequent evolution is determined by the opacity at each frequency, reaching flux maximum when the component becomes optically thin. For frequencies at which the source becomes optically thick at a given time, this results in a slower initial increase of the flux at lower frequencies, with the flux peak higher and occurring earlier at higher frequencies. Once the maximum flux is reached the light curves show an exponential decay, in agreement with the light curves of superluminal components observed in 3C 345 (Lobanov and Zensus 1997) and the radio flares in the accreting binary system GRS 1915+105 (Foster et al. 1996). Small fluctuations are present in the light curves, apparent at epochs 1.0 and 1.3 yrs. These correspond to changes in the brightness distribution of the component, when the peak of flux moves to the subsequent knot as the component evolves along the jet (see also Fig. 2).

4. CONCLUSIONS

The inclusion of the time-dependent simulations of the relativistic dynamics in the computation of the emission from relativistic jets should prove to be a powerful tool toward understanding the physics of superluminal sources. Besides revealing the existence of standing shocks and internal structure of moving components, our simulations show that a single perturbation can create several distinct features as it propagates downstream. Some indications of this effect are suggested by the C4 component in 3C 345 (see Lobanov and Zensus 1997). In addition, the interaction of moving components with standing shocks may result in a temporary 'drag' of the latter. An indication of this behavior has been detected in component K1 of 0735+178 by Gabuzda et al. (1994).

Acknowledgements

This research is supported in part by the Spanish DGICYT (PB94-1275, PB94-0973) and by NATO grant SA.5-2-05(CRG.961 228). J. L. Gómez and J. Mª. Martí gratefully acknowledge return grants from the Spanish Ministry of Education.

REFERENCES

Alberdi, A., Marcaide, J.M., Marscher, A.P., Zhang, Y.F., Elosegui, P., Gómez, J.L., & Shaffer, D.B., 1993, *Ap.J.*, **402**, 160.

Daly, R.A., & Marscher, A.P., 1988, *Ap.J.*, **334**, 539.

Duncan, G.C., & Hughes, P.A., 1994, *Ap.J.Lett.*, **436**, L119.

Foster, R.S., Waltman, E.B., Tavani, B., Harmon, B.A., Zhang, S.N., Paciesas, W.S., & Ghigo, F.D., 1996, *Ap.J.Lett.*, **467**, L81.

Gabuzda, D.C., Wardle, J.F.C., Roberts, D.H., Aller, M.F., & Aller M.H., 1994, *Ap.J.*, **435**, 128.

Gómez, J.L., Alberdi, A., & Marcaide, J.M., 1993, *A&A*, **274**, 55.

Gómez, J.L., Alberdi, A., Marcaide, J.M., Marscher, A.P., & Travis,

J.P., 1994, *A&A*, **292**, 33.

Gómez, J.L., Martí, J.Mª., Marscher, A.P., Ibáñez, J.Mª., & Marcaide, J.M., 1995, *Ap.J.Lett.*, **449**, L19.

Gómez, J.L., Martí, J.Mª., Marscher, A.P., & Ibáñez, J.Mª., 1996, in Blazar Continuum Variability, eds. H. R. Miller, J. R. Webb & J. C. Noble. ASP Conference Series, **110**, 242.

Gómez, J.L., Martí, J.Mª., Marscher, A.P., Ibáñez, J.Mª., & Alberdi, A., 1997, *Ap.J.Lett.*, in press.

Hughes, P.A., Aller, H.D., & Aller, M.F., 1989, *Ap.J.*, **341**, 54.

Hughes, P.A., Aller, H.D., & Aller, M.F., 1991, *Ap.J.*, **374**, 57.

Junor, W., & Biretta, J.A., 1995, *A.J.*, **109**, 500.

Komissarov, S.S., & Falle, S.A.E.G., 1996, in Energy Transport in Radio Galaxies and Quasars, eds. P. E. Hardee, A. H. Bridle, J. A. Zensus. ASP Conference Series, **100**, 173.

Lobanov, A.P., & Zensus, J.A., 1997, *Ap.J.*, submitted.

Marscher, A.P., & Gear, W.K., 1985, *Ap.J.*, **298**, 114.

Martí, J.Mª., Müller, E.,& Ibáñez, J.Mª, 1994, *A&A*, **281**, L9.

Martí, J.Mª., Müller, E., Font, J.A., & Ibáñez, J.Mª, 1995, *Ap.J.Lett.*, **448**, L105.

Martí, J.Mª., Müller, E., Font, J.A., Ibáñez, J.Mª, Marquina, A., 1997, *Ap.J.*, **479**, 151.

Mioduszewski, A.J., Hughes, P.A., & Duncan, G.C., 1997, *Ap.J.*, **476**, 649.

Pacholczyk, A.G., 1970, Radio Astrophysics, Freeman, San Francisco.

Tingay, S.J. et al., 1995, *Nature*, **374**, 141.

Wehrle, A.E. et al., 1997, in Blazar Continuum Variability, eds. H. R. Miller, J. R. Webb & J. C. Noble. ASP Conference Series, **110**, 430.

Wilson, M.M., & Scheuer, P.A.G., 1983, *MNRAS*, **205**, 449.

JETS IN AGN:
THE LARGE SCALE

X-RAY EMISSION AND CONFINEMENT OF JETS

E. TRUSSONI

Osservatorio Astronomico di Torino, Strada dell'Osservatorio 20, I-10025 Pino Torinese, Italy

The structure and evolution of radio galaxies strongly depend on their interaction with the external medium. This interaction is evident through the different peculiar morphologies observed in these objects (oscillations, head tails, hot spots, etc.). The environment is usually detected as X-ray emission from a medium embedding clusters and groups of galaxies, or surrounding isolated galaxies as hot coronae. The main properties of this plasma, deduced from theoretical models and satellite observations, are discussed in connection with their interaction with the radio sources. The major problem is whether or not this medium can confine jets and radio lobes, and this point will be discussed separately for the different classes of radio galaxies. We will see in particular that bright sources are in equilibrium with their environment, while faint objects appear underpressured. The consequences of these data are shortly discussed.

1. INTRODUCTION

The extended extragalactic radio sources are classified in the two classes *Fanaroff-Ryley I* (FRI) and *Fanaroff-Ryley II* (FRII), depending on their morphology and power. In the first class we find bright and symmetric radio sources with hot spots and narrow, mainly one-sided, radio jets. The FRI sources are weaker, with more peculiar structures: twin jets with large amplitude wiggles smoothly merging into extended lobes, no hot spots, more or less wide head tail structures.

The different morphologies observed in FRI and FRII objects can be due to intrinsic properties of the radio galaxies. The one-sidedness is likely to be related to the relativistic velocity of jets. Furthermore, wiggles and oscillations can be ascribed to precession effects in the nucleus where jets originate. However the interaction of the radio components (jets and lobes) with an external medium surely plays a main role in governing the structure of extended radio galaxies. For example, it is difficult to understand the morphology of head tail sources unless we assume that jets are bent by the motion of the parent galaxy through an external medium.

The presence of the intergalactic and intracluster gas has been confirmed by the discovery of extended clouds of hot plasma, emitting at X-ray energies, embedding clusters, groups and early type galaxies. Observations from satellites at these frequencies allowed to estimate several physical parameters of this gas (temperature, density, chemical composition), that are fundamental to understand its interaction with the radio components. Here we discuss and summarize the main features of this interaction starting from observational constraints and theoretical models, mainly addressing the problem of the confinement of radio jets.

2. JET - ENVIRONMENT INTERACTION

The presence of an intergalactic medium can affect the structure of jets in two main ways. A beam propagating through an ambient gas undergoes the effect of the ram pressure $P = \rho_{ext} v_h^2$, where v_h is the head's velocity and ρ_{ext} the density of the external gas. A shock is present in the contact region, where the jet material, crossed the shock, backflows creating a cocoon of processed plasma that surrounds the jet. The hot spots present in the radio lobes of FRII sources can be easily associated with these shocks, where effective acceleration of relativistic particles occurs (Muxlow and Garrington 1996, Lehay 1991, Williams 1991). The environment is also responsible for the structure of the head tail sources as due to the transverse propagation of the jets through the ambient plasma. It is possible to see that $\mathcal{R}_j \propto 1/v_{gal}^2$, where $\mathcal{R}_j$ is the curvature radius of the bent beam and v_{gal} the velocity of the parent galaxy with respect to the gas, while the jet Mach number $M^2 \approx \mathcal{R}_j/r_j$, where r_j is the jet radius. It is evident that the jet is more bent (small $\mathcal{R}_j$) the less supersonic is its speed and the larger is the velocity of the galaxy (Icke 1991).

The second major effect of an intergalactic gas is its role in the confinement of jets. Every outflow can appear collimated if it is expanding transversally with very small opening angle ξ. However the intrinsic properties are quite different whether or not the jet is freely expanding. In the former case we have, assuming the polytropic relation $P \propto \rho^\alpha$: $\xi \equiv \mathrm{d}\, r_j/\mathrm{d}\, R_j \approx 2/[(\gamma - 1)M_o]$, with $\rho_{axis} \propto R^{-2}$ and $P_{axis} \propto R^{-2\alpha}$, where γ is the ratio of the specific heats and M_o the initial Mach number. In a jet in equilibrium with its surrounding gas, the physical quantities have a different scaling: $A \propto r_j^2 \propto P^{-1/\alpha}, M \propto (P\,A)^{-1/2}$.

A jet can be considered confined depending on its structure and that of the external medium, in particular on the dynamical distance inside the beam, $l_j \equiv 2r_j M$, and the scale length of the outer atmosphere $l_a \equiv |P_a(\mathrm{d}P_a/\mathrm{d}R)^{-1}|$. If $l_j \ll l_a$, the beam is confined, in which case we have for the opening angle $\xi \ll 1/(\alpha M)$. For example, for an atmosphere with $P_a \propto R^{-a}$, which means $l_a \propto R$, since is $l_j \propto P^{-1/2}$ a jet is in equilibrium or not if $a < 2$ or $a > 2$, respectively. From the observations most jets have null or very small opening angle, implying confinement from the intergalactic gas. Furthermore, in beams showing variable widths along their axis, both regions of free expansion and recollimation are likely to be present, related to the particular structure of the environment.

The evolution of a confined jet depends on two important physical processes. First of all, the recollimation process is likely to work through the onset of lateral shocks, where acceleration of particles can occur. More important is the possible development of the well known Kelvin-Helmholtz instability at the boundary of an equilibrium jet. As largely shown in these last years, the onset of this instability can be related to several physical and morphological features observed in extended radio sources: peculiar structures (wiggles and blobs), formation of internal shocks, development of turbulence, entrainment of external gas, acceleration of relativistic particles. We do not comment here the details of this process (see e.g. Birkinshaw 1991, Ferrari et al. 1996, Chapter 15), but we discuss how the confinement conditions can be fulfilled, taking into account what we have learned on the extragalactic plasma from the X-ray observations.

3. EQUILIBRIUM CONDITIONS

A radio jet is in equilibrium with its surrounding medium if the internal (P_i) and external pressures (P_e) balance across the contact surface, $P_i = P_e$. From the observations the values of the two pressures are deduced separately from X-ray and radio data.

Internal pressure: Inside the jet it is assumed equipartition of energy between radiating particles and magnetic fields, in which case the energy density is minimum u_{min}. However, in the standard formulae providing u_{min} some further assumptions are necessary (Pacholczyk 1970):

1) The energy densities in relativistic electrons and heavy particles are equal $(k = 1)$;

2) The same volume is occupied by radiating electrons and magnetic field (filling factor $\Phi = 1$);

3) No thermal plasma is present;

4) There is no low energy tail in the population of the radiating electrons (cut off in the radio spectrum at 10 MHz).

External pressure: The pressure of the external medium is easily deduced if it is hot enough to emit at X-ray energies. Such a plasma has been discovered to be quite a common feature in single galaxies (galactic coronae) and in groups and clusters of galaxies. The main data on this environment are then obtained from satellite observations. In particular, from the spectral analysis we deduce the physical parameters (temperature, chemical composition and luminosity), while from the morphological data we get the brightness distribution, and thence the radial profile of the density and of the pressure. In this way we can compare locally the behavior of P_i and P_e as a function of the distance from the center of the galaxy.

4. GALACTIC HALOS AND INTRACLUSTER GAS

The main data on hot galactic coronae and intracluster gas have been obtained by satellites operating in the soft - medium X-ray energy band ($\approx 0.1 - 10$ keV), and in the following we summarise their main properties (Forman et al. 1985, Sarazin 1986, Canizares et al. 1987, Fabbiano 1989).

4.1. General structure

With very simple arguments we can see that a gas in the potential well of a cluster must be hot. Let us consider the galaxy system as a gas with temperature T_{cl}, with σ_r the value of the radial velocity dispersion, then $T_{cl} \propto \sigma_r^2$. If the system is embedded in a gas cloud with temperature $T_g \approx T_{cl}$ we have (k is the Boltzmann constant and m_p the mass of the proton):

$$\frac{k}{m_p} T_g \sim \sigma_r^2 \quad ,$$

and therefore

$$T_g \sim 10^8 \left(\frac{\sigma_r}{10^3\,\text{km/s}}\right)^2 \text{K}$$

If the cluster is virialized we can estimate immediately the mass of the hot gas, assumed of size R (G is the gravitational constant):

$$\frac{M}{M_\odot} \sim \frac{3}{G} R\sigma_r^2 \sim 6 \times 10^{13} \left(\frac{R}{10^2\,\text{Kpc}}\right)\left(\frac{\sigma_r}{10^3\,\text{km/s}}\right)^2$$

A similar argument basically holds for a hot galactic corona where $T_{cl} \to T_*$, with T_* 'temperature' of the star system.

The detailed properties of the intracluster and interstellar gas depend on the dynamics of the galaxy and of the cluster. The structure and evolution of these systems has been one of the major topics of the extragalactic astrophysics in these last two decades (see e.g. Cavaliere and Fusco-Femiano 1981, Sarazin 1986).

If we consider the stars system in a galaxy as an isothermal gas in its gravitational potential well, then the star density distribution with R follows the well known *King*'s law:

$$n_*(R) = \frac{n_{*,c}}{[1 + (R/R_c)^2]^{3/2}}$$

where $n_{*,c}$ is the central stellar density and R_c the core radius. From X-ray data it results that in most cases the brightness profile is (Forman et al. 1985):

$$S(R) = S_0 \left[1 + (R/R_c)^2\right]^{-3\beta+1/2}$$

If we assume that this emission comes from an optically thin thermal emitting plasma, then its density distribution agrees with the King's law, but with the exponent $3/2 \to 3\beta/2$, with β defined as the ratio between the energy density of the stars and of the hot gas: $m_p\sigma_r^2/2kT_g$.

A similar behavior has been found also in groups and clusters of galaxies. Spectroscopic observations confirm that the X-ray emission comes from a plasma with temperatures $\sim 10^7$ K and metal abundances $\mathcal{M}$ from few per cent to half the standard cosmic values. In Table 1 we summarize the main average properties of the intergalactic and coronal gas.

TABLE 1: Physical parameters of the hot plasma

	Clusters	Galactic coronae
R_c (Kpc)	300 - 500	2 - 10
n_c (cm^{-3})	$\sim 10^{-4}$	$\sim 10^{-2}$
β	0.4 - 0.6	0.6 - 0.8
T(K)	$2 - 8 \times 10^7$	$0.5 - 1.5 \times 10^7$
$\mathcal{M}$	~ 0.1	~ 0.5
L_X (erg s^{-1})	10^{44-45}	10^{41-42}
M_{gas} $(M_\odot)$	10^{14-15}	10^{9-10}

4.2. Cooling flows

In several clusters a drop of the temperature is found in their central region. This is expected because in those zones the cooling time of the hot gas, due to the radiative losses, is usually shorter than the Hubble time. Accordingly we expect that the pressure drops, driving the infall of the gas towards the center of the cluster or of the galaxy. However a real inwards motion of the plasma in its inner region as not been detected yet.

4.3. Dark matter

It is well known from their curve of rotation that in several galaxies only a small fraction of the total mass is in the stars. This implies the presence of dark halos of mass $M_d \gg M_*$ with density profile (Binney and Tremaine 1987):

$$n_d(R) = \frac{n_{d,c}}{1 + (R/R_{c,d})^2}$$

X-ray data confirm that most of the mass in galaxies and clusters is not visible. If the hot gas is in hydrostatic equilibrium and isothermal, the underlying binding mass is:

$$\frac{M_{grav}}{M_\odot} = 1.1 \times 10^{11} \beta \left(\frac{T}{10^7\,\text{K}}\right) \left(\frac{R_c}{1\,\text{kpc}}\right) \frac{(R/R_c)^3}{1 + (R/R_c)^2}$$

We see that in most clusters and groups is $M_{grav} \gg M_{gal} + M_{gas}$, where M_{gal} is the total mass of the galaxies and M_{gas} that of the hot intracluster medium, deduced from the X-ray observations.

In several early type galaxies associated with coronae it is found $M_{grav}/L_B \gg M_*/L_B \approx 6$, the value expected in these objects (Forman et al. 1985; M_* and L_B are given in solar units).

4.4. Origin of the hot intergalactic plasma

The gas in clusters could be primordial, however the presence of a sensible fraction of heavy elements implies that it has been affected by processes of stellar evolution. Furthermore it is always $\beta < 1$, i.e. the gas temperature is higher than the temperature of the stars (or galaxies), so that the plasma must have been heated.

It is generally believed (David et al. 1990, 1991), as confirmed by numerical simulations, that at the beginning of their age the forming galaxies undergo a phase of strong rate of star formation with supernova explosions. This drives a strong galactic wind that heats the primordial gas, providing also enrichment of heavy elements. Thence a quiescent phase follows, in which the galactic winds stops and an inner cooling flow sets up. Finally, the gas can reach an almost stationary configuration in the galactic or cluster potential well. The rate of supernovae explosions is much lower, but the gas can be further heated by the mass loss from stars.

5. CONFINEMENT OF RADIO SOURCES

Following Sec. 3, we now discuss the possibilities of confinement of radio jets according to their properties at radio frequencies.

5.1. Low brightness Radio galaxies (FR I)

As pointed out in the Introduction, in this class of objects we find several peculiar morphologies that suggest a strong interaction with the ambient medium: head tail, wiggles, tunnels. Morganti et al. (1988) analyzed a sample of radio galaxies embedded in X-ray emitting plasma, and found in most cases $P_i < P_e$. This behavior, further confirmed in recent years by more detailed observations, holds for clusters or isolated galaxies (see Feretti et al. 1995, 1997, Massaglia et al. 1996, Trussoni et al. 1997). We see in particular that the discrepancy can be more than one order of magnitude in groups or clusters.

Some known examples with opposite trend are the jets in M87 and NGC 6251, that appear overpressured (Biretta 1996). In the second object the jet appears quite unperturbed, showing slight interaction with the external gas. The beam in M87 is quite peculiar, showing a limb brightening that could be related to a hollow structure.

It is believed that jets are actually in equilibrium with their environment, and that the observed imbalance must be related to some 'external' bias. We must remind first that projection effects and the clumpy distribution of the· external gas could affect the data. However it is likely that the discrepancy between the inner and the outer pressure originates from the several assumptions followed to evaluate P_e (see Sec. 3). The plasma could be out of equipartition, with low filling factor ($\Phi \gg 1$) and with a contribution to its internal energy from other particles than electrons ($\kappa \gg 1$). Furthermore, thermal gas may be present, and the electron distribution could have a low energy tail, with no cut off of radio emission at 10 MHz.

5.2. Cooling flows

Most of the radio sources associated with cooling flows belong to the FRI class and appear quite bright and of small extension. They usually appear with two typical morphologies (Sarazin 1996):

a) Twin lobes. These objects, associated with jets, show a sort of 'spatial' anticorrelation: the region of X-ray and radio emission are separate, as though the radio component 'digged' its way through the environment. This is confirmed by radio measurements of Faraday rotation, likely related to the external gas swept by the radio blob. In these sources is usually $P_e \approx P_i$

b) Amorphous. For such morphologies the regions of radio and X-ray emission overlap, suggesting a strong mixing between the two plasma components. This is consistent with depolarization measurements at radio frequencies. In such a case it is found $P_e \gg P_i$.

From the above considerations it is obvious the strong interaction between the radio component and the hot plasma in a cooling flow. In particular we can argue that an emerging jet cannot propagate far from its parent galaxy, but dissipates most of its energy on small scales interacting with the flow, producing powerful, but small radio sources.

Cygn A is in contrast with this picture: in fact it belongs to the FR II class, showing no depolarization or Faraday rotation. The possible explanation for this behavior is that the X-ray emission is of non-thermal origin (Harris et al. 1994).

5.3. Powerful Radio Galaxies (FR II)

The main properties of a complete sample of FR II radio sources and their environments has been recently analyzed by Wan and Daly (1996). First of all, the size and pressure P_i of the FR II appear independent on the redshift and on the 'habitat', i.e. whether or not belong to a cluster, and which type of cluster. Furthermore, for low z, FR II sources are found in faint X-ray cluster, where the pressure of the external gas is much lower than in bright X-ray clusters (no certain data are available for high z).

The most important result with respect to FRI objects is that basically in all the sources is $P_e \sim P_i$, within a factor ≈ 3. This quite homogeneous behavior implies that also the evolution of FR II sources is strongly ruled by the environment. Namely the sources always seem to reach equilibrium with their surrounding gas; furthermore they can expand only if the external pressure is not too high (i. e. in faint X ray clusters).

6. SUMMARY AND CONCLUSIONS

Radio and X-ray data confirm that in all classes of radio sources jets in equilibrium with the environment should be a common feature. In FR I objects the confinement is likely to trigger the Kelvin-Helmholtz instability, that in its non linear phase first develops internal shocks, leading finally to strong mixing between the jet's material and external gas (Ferrari et al. 1996). This evolution does not critically depend whether or not these sources belong to a cluster or group, apart from the case of head tail structures. In FR II radio sources the picture is different: the instability appears much less efficient, probably due to the relativistic velocities, while the hot pressure in bright cluster seems to prevent the formation of an extended object. A similar behavior is found in cooling flows: the kinetic energy of the jet can either be dissipated close to the center or can dig short tunnels through the ambient gas. Numerical simulations have shown that only jets with very high Mach numbers (> 20) can propagate at large distances (Loken et al. 1993).

We can conclude by noticing that several properties of radio jets have been found on smaller scale in stellar jets. In particular the structure of collimated outflows in Young Stellar Objects can be interpreted in terms of the non-linear evolution of the Kelvin-Helmholtz instability (Ray 1996), implying the presence of a confining environment. On the other hand it is interesting to notice that some peculiar morphologies of radio galaxies (e.g. head tail) are not found in these galactic objects. Therefore more detailed data on stellar jets will be helpful to further understand the dynamical interaction between the various classes of radio sources with the external gas.

REFERENCES

Binney, J., & Tremaine, S., 1987, *Galactic Dynamics*, Princeton Univ. Press.

Biretta, J.A., 1996, in *Solar and Astrophysical MHD Flows*, K.Tsinganos (ed.), Kluwer Academic Publishers, 367.

Birkinshaw, M., 1991, in *Beams and jets in astrophysics*, P.E. Hughes (ed.), Cambridge University Press, 278.

Canizares, C.R., Fabbiano, G., & Trinchieri, G., 1987, *Ap.J.*, **312**, 503.

Cavaliere, A., & Fusco-Femiano, R., 1981, *A&A*, **100**, 194.

David, L.P., Forman, W., & Jones, C., 1990, *Ap.J.*, **359**, 29.

David, L.P., Forman, W., & Jones, C., 1991, *Ap.J.*, **369**, 121.

Fabbiano, G., 1989, *Ann. Rev. Astr. Astr.*, **27**, 87.

Feretti, L., Fanti, R., Parma, P., Massaglia, S., Trussoni, E., & Brinkmann, W., 1995, *A&A*, **298**, 699.

Feretti, L., Böhringer, H., Giovannini, G., & Neumann, D., 1997, *A&A*, **317**, 432.

Ferrari, A., Massaglia, S., Bodo, G., & Rossi, P., 1996, in *Solar and Astrophysical MHD Flows*, K.Tsinganos (ed.), Kluwer Academic Publishers, 607.

Forman, W., Jones, C., & Tucker, W., 1985, *Ap.J.*, **293**, 102.

Harris, D.E., Carilli, C.L., & Perley, R.A., 1994, *Nature*, **367**, 713.

Icke, V., 1991, in *Beams and jets in astrophysics*, P.E. Hughes (ed.), Cambridge University Press, 232.

Leahy, J.P., 1991, in *Beams and jets in astrophysics*, P.E. Hughes (ed.), Cambridge University Press, 100.

Loken, C., Burns, J.O., Norman, M.L., & Clarke, D.A., 1993, *Ap.J.*, **417**, 515.

Massaglia, S., Trussoni, E., Caucino, S., Fanti, R., Feretti, L., Parma, P., & Brinkmann, W., 1996, *A&A*, **309**, 75.

Morganti, R., Fanti, R., Gioia, I.M., Harris, D.E., Parma, P., & de Ruiter, H., 1988, *A&A*, **189**, 11.

Muxlow, T.W.B., & Garrington, T., 1991, in Beams and jets in astrophysics, P.E. Hughes (ed.), Cambridge University Press, 52.

Pacholczyk, A.G., 1970, *Radio Astrophysics*, S. Francisco, Freeman.

Ray, T.P., 1996, in *Solar and Astrophysical MHD Flows*, K.Tsinganos (ed.), Kluwer Academic Publishers, 539.

Sarazin, C.L., 1986, *Rev. Mod. Phys.*, **58**, 1.

Sarazin, C.L., 1996, in *Röntgenstrahlung from the Universe*, H.U. Zimmermann, J.E. Trümper and H. Yorke (eds.), MPE Rep. 263, 561.

Trussoni, E., Massaglia, S., Ferrari, R., Fanti, R., Feretti, L., Parma, P., & Brinkmann, W., 1997, *A&A*, in press.

Wan, L., & Daly, R.A., 1996, *Ap.J.*, **467**, 145.

Williams, A.G., 1991, in *Beams and jets in astrophysics*, P.E. Hughes (ed.), Cambridge University Press, 342.

CHAPTER 11

ASYMMETRY IN EXTRA-GALACTIC RADIO SOURCES: ABERRATED, RADIATIVE, INITIAL?

D. FRAIX-BURNET

*Laboratoire d'Astrophysique de l'Observatoire de Grenoble,
BP 53, F-38041 Grenoble Cédex 9, France*

Among the asymmetries found in radiosources, one-sidedness of extragalactic jets is the most spectacular and may have three origins. It can be explained by an aberration effect from two intrinsically symmetric jets, or by the fact that the two opposed jets have different radiative or even physical properties. I will review some of the evidences that make the first explanation not sufficient in itself. Consequently, at least in a few cases, the two opposed jets are intrinsically different. The asymmetry probably comes from different radiative behavior, but different physical properties are plausible. The role of the environment is certainly important but not fully clear yet.

1. INTRODUCTION

Extragalactic radiosources very often show structural and physical asymmetries between the two sides. The lobes generally have different shapes, sizes, spectral indices, polarization and the lengths of the arms are also different. The environment can be invoked, but certainly there are intrinsic physical properties as well. The most spectacular asymmetry remains the one-sidedness of jets observed in roughly 30% of the radiosources.

Even if a *flip-flop* mechanism cannot be entirely ruled out, it is generally believed that the two opposed jets are ejected at the same time. For simplicity and lack of observational constraints, it is also generally assumed that they are ejected with the same density, energy and magnetic field, so that any brightness difference can only be explained by an aberration effect. However, the three possible causes for one-sidedness can be classified following a hierarchy in the complexity of the physics:

1) initial properties: *symmetric*; radiation: *symmetric*; relativistic aberration: *necessary*.
2) initial properties: *symmetric*; radiation: *asymmetric*; relativistic aberration: *not necessary*.
3) initial properties: *asymmetric*; radiation: *symmetric or asymmetric*; relativistic aberration: *not necessary*.

Of course they are not exclusive to each other and certainly are all found in nature. However, one might hope each of them to be dominant in different sources so that sometimes the three can be disentangled. The first explanation is clearly not concerned very much with the physics of the radiation of jets. However, the one-sidedness necessarily is tightly linked to the particle acceleration and radiation processes, and significantly depends on local conditions in and around a given jet. It also probably implies an asymmetric interstellar medium (see below). The intrinsic asymmetry from the origin has been recently put forward and shown to be theoretically possible (Wang, Sulkanen and Lovelace 1992; Chagelishvili, Bodo and Trussoni 1996). A particular configuration of the magnetic field is required: is this caused by a large scale asymmetry in the ISM that influences the magnetic field in the accretion process? This initial asymmetry will not be considered in more detail here. It should be reminded as a plausible explanation but requires the first two explanations for one-sidedness to be ruled out observationally.

In the eighties, a lot of theoreticians were interested in synchrotron radiation and in situ particle acceleration. The question has been somewhat forgotten, either because it has been considered to be resolved, or more probably because no more progress was possible from the observational constraints. I think that considering intrinsic asymmetry in jets is a new way of investigating the question of what we exactly see in a jet and how particles are accelerated.

In this paper, a two-fluid approach is assumed (Sol, Pelletier and Asséo 1989; Achatz, Lesch and Schlickeiser 1990; Fraix-Burnet and Pelletier 1991; Pelletier and Sol 1992). Indeed, the jets are bound to have two plasmas, one dynamical component (electrons/protons) extracted from the accretion disc and digging its way through the ISM, and a second one (electrons/positrons) responsible for synchrotron radiation. The first one (the 'thermal' plasma) is the support of magnetohydrodynamic turbulences that are necessary to accelerate the radiating particles. Only the second plasma (indeed the only one we see) needs to be relativistic as is shown for instance in Despringre and Fraix-Burnet (1997).

2. ORIGIN OF THE SYNCHROTRON RADIATION

From lifetime considerations, particles radiating in synchrotron have to be accelerated in the jet. This is particularly true in long radio jets and in optical jets. The constraint of producing a power-law spectrum with a quasi-universal spectral index, as more or less observed, has made the first-order Fermi acceleration very successful and popular. It requires a shock front, magnetic turbulences (Alfvén waves) and of course electrons of sufficient initial energy to enter the acceleration process (the so-called problem of the injection energy). The first two ingredients are borne by the thermal plasma, and the particles above the threshold energy define the second plasma, i.e. the relativistic component. This acceleration mechanism encounters some difficulties now, because no strong shock fronts are seen in very high resolution maps. In addition, the amount of energy available in the magnetic turbulences is unknown, so that it is still not possible to guarantee theoretically that this process is able to produce the observed synchrotron flux. Finally, the well-known problem of the injection energy (the electrons of the thermal plasma need a first acceleration of unknown origin) is still open. The second-order Fermi acceleration (without shock front) could work indeed, if one considers that the power-law shape of the spectrum could be due to spatial averaging of different spectra or of a distribution of magnetic fields (see Eilek and Arendt 1996).

Whatsoever, magnetic turbulences seem compulsory, and they can only be produced by the interaction of the jet with the interstellar medium. The boundary layer, produced by this interaction, mainly depends on their relative density ratio (e.g. Norman, Winkler and Smarr 1984) and determines the spectrum of turbulences. The entire cascade that leads to synchrotron radiation was developed in the early eighties (Eilek 1979, 1982; Ferrari, Trussoni and Zaninetti 1979, 1980, 1981; Ferrari, Massaglia and Trussoni 1982; Bicknell and Melrose 1982, see Fig. 1):

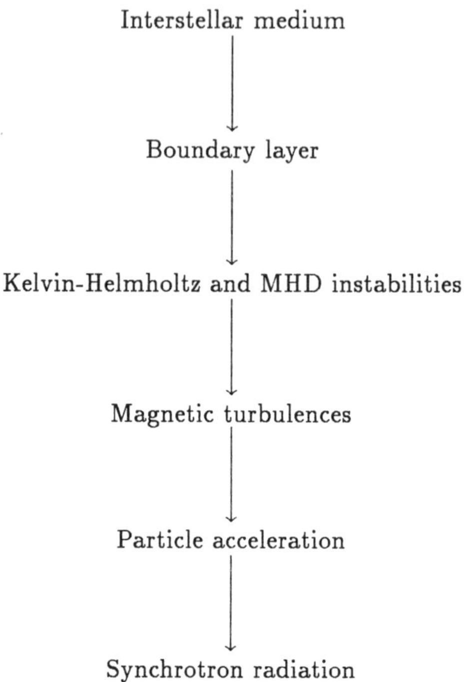

FIGURE 1. Cascade leading to synchrotron radiation in radio sources.

Fraix-Burnet (1992) confronted this scheme with observational results. Clearly one-sidedness is not due solely to an aberration effect, and both observations and theory can be reconciled by realizing that the synchrotron radiation is more important when the interaction of the jet with the ISM is weak. The non-thermal radiation is here determined by the spectrum of the (magnetic) turbulences and is very sensitive to their small scales that accelerate the particles. Too strong a jet/ISM interaction will produce large scales of turbulences and consequently heating (hence thermal emission) and entrainment, but inefficient particle acceleration and very little synchrotron radiation. The consequence of one-sidedness is that the ISM must be intrinsically asymmetric on the two sides of some galaxies if the above theoretical scheme is correct. This is consistent with the still relatively weak observational constraints.

Other acceleration mechanisms could be discovered, but it seems very improbable that local conditions (mainly imposed by the interaction of the jet with the ISM) do not play an essential role in the radiation process.

3. INTRINSIC DIFFERENCES IN RADIATION PROPERTIES

The radiation and the loss of kinetic energy of the jet are thus linked through the jet/ISM interaction. A correlation is then expected between the radiation of a jet and physical properties of a lobe in which the jet disperses its energy. This is indeed observed in the triple correlation between jet visibility, depolarization and spectral index of the corresponding lobe. This is a kind of 'generalized' Laing-Garrington effect (Fraix-Burnet 1992) that includes an intrinsic property of the physics of the lobes, that is the spectral index. This correlation also applies to un-jetted sources (Liu and Pooley 1991). This is a clear indication that the radiative asymmetry between two opposed jets is intrinsic and not due solely to Doppler aberration.

A few one-sided sources cannot be explained by such a relativistic effect: M 87 (Reid et al. 1989), NGC 6251 (Jones and Wehrle 1994), Cyg A (Carilli and Bartel 1991) and 1928+738 (Hummel et al. 1992). They are among the best studied radiosources and thus represent a good opportunity to better understand the physical processes that lead to synchrotron radiation.

A still stronger case has been found by Fraix-Burnet (1997) on 3C 66B. This source is unique because it has a double-sided radio jet and a one-sided optical jet. It clearly appears that the optical counterpart of the radio counter-jet has not the brightness expected from the relativistic aberration interpretation of the radio brightness asymmetry. This is the first clear case for which the synchrotron spectrum is shown to be intrinsically different in the two opposed jets.

If one-sidedness is only due to a relativistic aberration, one would expect the lobe on the visible jet side to be systematically longer (i.e. farther away from the nucleus) in one-sided sources. This is because of propagation time effect (the Ryle effect) and is modelised by Scheuer (1995). However, he finds in his sample that this is true in only 60% of the cases, so that intrinsic effects should dominate and that intrinsic asymmetries in the ISM are also certainly present. This result is confirmed on an essentially different sample by Kotanyi and Fraix-Burnet (in preparation).

4. CONCLUSION

Some definitive results can now be drawn about asymmetries in radiosources:

1) intrinsic radiative asymmetries are present in jets;

2) relativistic aberration is probably not a dominant effect in kpc-scale jets;

3) relativistic bulk motions are not required and a two-fluid concept is more suitable;

4) strong shocks are not seen in jets, implying either a lot of small (and weak?) shocks, or a particle acceleration mechanism without shocks fronts;

5) interstellar medium asymmetries are also present, even though it is still difficult to quantify them and directly relate with jet radiation;

6) radiosources arm-length asymmetries are mainly intrinsic and imply an asymmetric ambient (interstellar and/or intergalactic) medium.

Disentangling aberration from intrinsic effects is certainly not obvious, except in rare cases such as 3C 66B (Fraix-Burnet 1997). Orientation and velocity constraints based on asymmetry should thus be considered only cautiously.

Some questions remains open:

7) the quasi permanence of the visibility or non-visibility of a jet all over its length seems to be in contradiction with the local nature of the generation of radiation via turbulences from the interaction of the jet with the interstellar medium. This point might be related with entrainment considerations.

8) what is the ISM component which is the most important to generate turbulences: the hot or cold gas, or could some individual heavy cloud determines the boundary layer over tens of kpc downward?

9) there is a strong correlation in one-sided sources between the direction of the pc-scale and the kpc-scale jets. Is the ISM asymmetry found at these two scales? Could this be explained by the whole accretion process that feeds the monster in the nucleus of a galaxy?

10) since no strong shocks are seen in jets, one should consider stochastic acceleration as more probable. It might also be worthwhile thinking to something else to accelerate relativistic particles that avoids the still annoying injection energy question.

11) the quantitative estimation of the amount of the jet kinetic energy going into magnetic turbulences has never been performed and requires non-linear MHD calculations. It is however the only means to know whether stochastic acceleration is quantitatively plausible in extragalactic jets.

All this shows that the synchrotron radiation from extragalactic jets is far from being understood. With the new approach which considers that relativistic aberration might not be the dominant effect and that intrinsic differences should exist, the study of one-sidedness and asymmetry brings a new insight into the physics of jets. As a last point, it is important to note that questions 7) and 9) could be simply explained if the initial properties of the two opposed jets are different.

REFERENCES

Achatz, U., Lesch, H., & Schlickeiser, R., 1990, *A&A*, **233**, 391.

Bicknell, G.V., & Melrose, D.B., 1982, *Ap.J.*, **262**, 511.

Carilli, C.L., & Bartel, N., 1991, *A.J.*, **102**, 1691.

Chagelishvili, G.D., Bodo, G., & Trussoni, E., 1996, *A&A*, **306**, 329.

Despringre, V., & Fraix-Burnet, D., 1997, *A&A*, in press.

Eilek, J.A., & 1979, *Ap.J.*, **230**, 373.

Eilek, J.A., 1982, *Ap.J.*, **254**, 472.

Eilek, J.A., & Arendt, P.N., 1996, *Ap.J.*, **457**, 150.

Ferrari, A., Trussoni, E., & Zaninetti, L., 1979, *A&A*, **79**, 190.

Ferrari, A., Trussoni, E., & Zaninetti, L., 1980, *MNRAS*, **193**, 469.

Ferrari, A., Trussoni, E., & Zaninetti, L., 1981, *MNRAS*, **196**, 1051.

Ferrari, A., Massaglia, S., & Trussoni, E., 1982, *MNRAS*, **198**, 1065.

Fraix-Burnet, D., 1992, *A&A*, **259**, 445.

Fraix-Burnet, D., 1997, *MNRAS*, **284**, 911.

Fraix-Burnet, D., & Pelletier G., 1991, *Ap.J.*, **367**, 86.

Hummel, C.A., Schalinski, C.J., Krichbaum, T.P., Rioja, M.J., Quirrenbach, A., Witzel, A., Muxlow, T.W.B., Johnston, K.J., Matveyenko, L.I., & Shevchenko, A., 1992, *A&A*, **257**, 489.

Jones, D.L., & Wehrle, A.E., 1994, *Ap.J.*, **427**, 221.

Liu, R., & Pooley, G., 1991, *MNRAS*, **253**, 669.

Norman, M.L., Winkler, K.-H.A., & Smarr, L.L., 1984, in *Physics of energy transport in extragalactic radio sources*, eds. A.H. Bridle & J.A. Eilek, NRAO, p. 150.

Pelletier, G., & Sol, H., 1992, *MNRAS*, **254**, 635.

Reid, M.J., Biretta, J.A., Junor, W., Muxlow, T.W.B., & Spencer, R.E., 1989, *Ap.J.*, **336**, 112.

Scheuer, P.A.G., 1995, *MNRAS*, **277**, 331.

Sol, H., Pelletier, G., & Asséo, E., 1989, *MNRAS*, **237**, 411.

Wang, J.C.L., Sulkanen, M.E., & Lovelace, R.V.E., 1992, *Ap.J.*, **390**, 46.

CHAPTER 12

JET-CLOUD INTERACTIONS IN SEYFERT GALAXIES

A. CAPETTI

Scuola Internazionale di Studi Superiori Avanzati,

Via Beirut 2-4, I-34014 Trieste, Italy

Recent HST observations showed that the Narrow Line Regions (NLR) of Seyfert galaxies originate and are governed by the interaction between jets and interstellar clouds. These results change dramatically our view on the NLR in Seyfert galaxies, and more generally in Active Galactic Nuclei, and open a different perspective for the study of both jets and emission line regions.

1. INTRODUCTION

The study of the emission line regions of Active Galactic Nuclei offers a unique tool to better understand the properties of their nuclear outflows and their interaction with the surrounding medium. In particular in this contribution I will produce clear evidence that the properties of the Narrow Line Regions (NLR) in Seyfert galaxies are dominated by the presence of radio emitting outflows.

Two mechanisms have been proposed for the line and the continuum emission of NLR. In the first of these, the nuclear radiation field dominates the energetics of the NLR. The gas producing the narrow lines is photoionized by the nuclear emission (e.g. Ferland and Osterbrock, 1986) which is also scattered into our line of sight. The evidence for illumination from a (anisotropic) radiation field is primarily based on the observation of scattered broad lines in Seyfert 2 galaxies (Antonucci and Miller, 1985), the discovery of the Extended

Narrow Line Region (ENLR) from ground based observations (Unger et al., 1987) and ionization cones co-spatial with the NLR (Wilson and Tsvetanov, 1994 and reference therein). While this 'standard' scenario appears to be overall in good agreement with observations, particularly for its ability of reproducing emission line ratios, it is not a predictive model with respect to other issues, such as the origin or the velocity field of the emitting clouds.

Ground based studies have shown that the NLR is generally co-spatial with the radio emission (Wilson and Ulvestad 1983) and its kinematic displays signs of the effect of the interaction with the ejected radio plasma (Meaburn and Pedlar 1986; Baldwin, Wilson and Whittle 1987). Haniff, Wilson and Ward (1988) found that on a sample of 10 Seyfert galaxies which show an extended structure at radio wavelength, the [O III] emission is always aligned within a few degrees to the radio axis, suggesting a connection between the relativistic and thermal gas. The association between the radio emitting components and the NLR prompted Pedlar et al. (1989) and Taylor et al. (1992) to propose the alternative model in which the structure of the NLR is dominated by the compression and heating of the interstellar gas generated by shock waves formed by the supersonic radio ejecta. The shock waves can also be responsible for the ionization of the thermal gas (Axon et al., 1994 and references therein) and might create significant continuum emission. As Sutherland, Bicknell and Dopita (1993) have argued, shock waves might create significant continuum emission, which might be important in ionizing the NLR. Therefore, the jet/cloud interaction model makes clear predictions on the origin, dynamics and ionization structure of the NLR, which can be tested with observations.

It is therefore of great interest to establish the relation between the radio and optical emission line regions in Seyfert galaxies. This is now possible with *Hubble Space Telescope* observations which spatially resolve the NLR in nearby Seyfert galaxies and allow us to compare their structure with the radio morphology. With this aim a project of HST imaging and spectroscopy of Seyfert galaxies was undertaken.

In this contribution I will focus on the results obtained from the observations of NGC 1068, which is one of the closest Seyfert galaxies and it harbors a bright radio source, with a prominent radio-jet which terminates in an extended radio-lobe. It is therefore an ideal candidate for a high spatial resolution HST study of its NLR and for a comparison with its radio properties. However, it must be emphasized that analogous results have been obtained for essentially all Seyfert galaxies studied with HST in which a radio outflow are present.

Throughout this paper a distance to NGC 1068 of 14.4 Mpc (Tully 1988) is adopted, where $1''$ corresponds to 72 pc.

2. THE HST IMAGING PROJECT

Figure 1 shows the radio contours of NGC 1068 from the 6 cm MER-LIN radio image of Muxlow et al. (1996) overlaid onto the [O III] $\lambda5007$ emission line image from Macchetto et al. (1994). The regis-tration between the two images is based on the absolute astrometry derived by Capetti, Macchetto and Lattanzi (1997) which is accu-rate to $0''.1$. There is a clear connection between radio and optical emission: the line-emission originates essentially in a series of knots located along the edges of the radio jet.

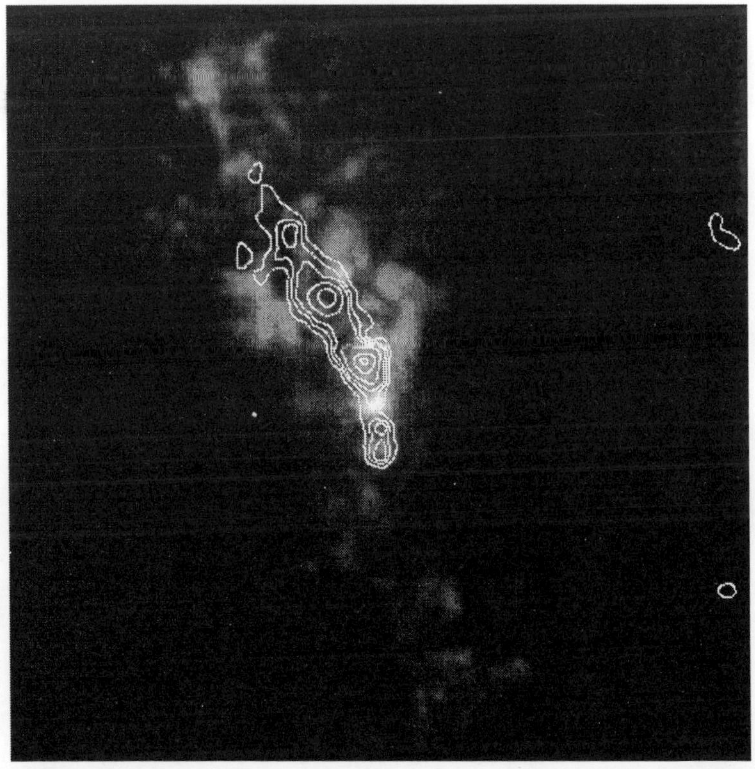

FIGURE 1: Contours of the 6 cm MERLIN radio image of Muxlow et al. (1996) overlaid onto the [O III] emission line image from Macchetto et al. (1994). Note the connection between radio and line emission. The field of view is $4'' \times 4''$.

The density of individual knots can be derived from the [S II] doublets in archival FOS/HST spectra and ranges between $10^{4.0}$ cm^{-3} and $10^{4.5}$ cm^{-3} depending weakly on the adopted temperature. In comparison the average density obtained from ground based long-slit spectroscopy varies from $10^{3.6}$ cm^{-3} to $10^{3.0}$ cm^{-3} between 0″0.5 and 1″0.5 from the nucleus. These estimates confirm the visual impression that the brightest knots are high density condensations within the NLR.

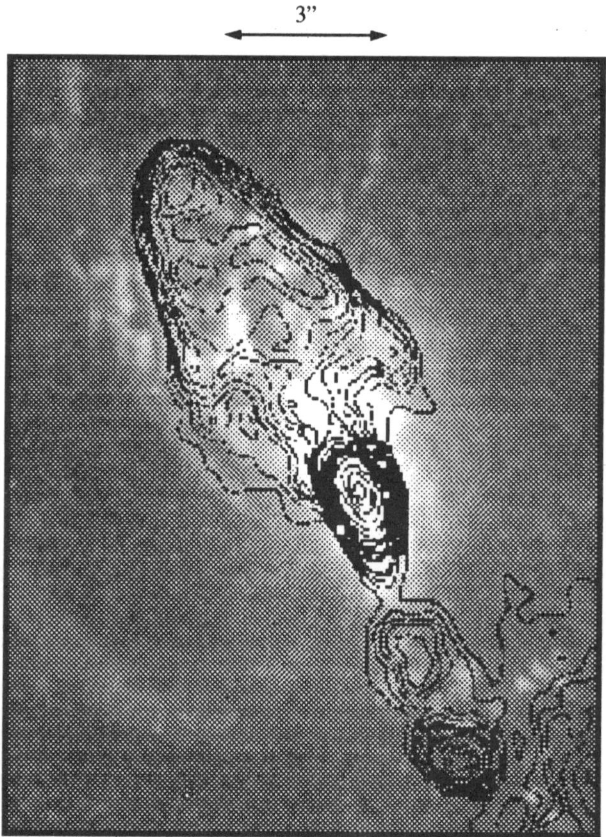

FIGURE 2: Radio contours of NGC 1068 from Wilson and Ulvestad (1983) superposed to a composite emission-line image obtained co-adding the [O III] and the Hα + [N II]λ6584 images, in order to show more clearly both the high and low ionization regions.

It therefore appears that the radio-jet is sweeping and shocking the external medium, effectively clearing a channel through the nuclear regions. The increase in the density due to the shocks causes the line-emission (which depends on the density squared) to be highly enhanced in the region where this interactions occurs. Similar results have been obtained from the observations of several other Seyfert with radio jets (e.g. Mrk 3, Mrk 348, Mrk 6, Mrk 1066) in which the NLR appears jet-like and is spatially coincident with the radio jet (Capetti et al. 1995a, 1995b, 1996, Bower et al. 1996).

The radio emission in NGC 1068 extends over $\sim 15''$, with two radio lobes forming a bipolar structure. Fig. 2 shows the large scale radio contour image from Wilson and Ulvestad (1983) overlaid on a composite emission-line image, obtained by co-adding the [O III] and the Hα+[NII] images from Capetti, Axon and Macchetto (1997).

This overlay reveals a strong correspondence between the emission line filament system and the Northern radio-lobe. The filaments are essentially confined to within the radio-emission and the sharp edges of the radio lobe are bracketed by two bright emission features. The northern edge of the lobe is clearly marked by a curved structure in emission-line. Outside the radio lobe the emission-line surface brightness drops dramatically. This confirms what is observed in other Seyfert galaxies in which a radio lobe is present (e.g. Mrk 573, Mrk 78, IRAS 0421+045 and IRAS 1105-035; Capetti et al. 1996, Axon et al. 1997a) in which the NLR appears to be bow-shock-like. Again, this structure can be explained by the sweeping-up of gas by the expanding and advancing radio lobes.

3. DYNAMICS OF THE NLR

Overall these results present dramatic confirmation of what was inferred from ground based observations, i.e. that the radio jets play a fundamental role in driving and forming the NLR. The dynamics of the NLR of NGC 1068 offer a crucial confirmation of this model. In fact we expect that the jet/cloud interaction will manifest itself in a clear dynamical signature.

The velocity field observed in the inner region of Seyfert galaxies is particularly complex, with line profiles significantly broader that the typical velocities inferred from stellar rotation curves and thus requiring an additional acceleration process other than gravity. This complexity is almost certainly to be ascribed also to the spatial confusion caused by seeing effects which mixes kinematic components from different radii. This makes the interpretation of the observed line profiles particularly difficult. However, the region where the jet interaction is occurring can be now spatially resolved with HST. For this reason long–slit spectroscopic observations have been obtained with the Faint Object Camera f/48 spectrograph at a spatial resolution of $0''.0287$ per pixel (Axon et al. 1997b).

The emission line region around the jet of NGC 1068 is broken up into a series of highly blue-shifted knots. Around $\pm 0.5''$ of the jet the emission lines are split into two velocity systems separated by 1500 km sec^{-1}. Outside this region the gas motions follow the overall pattern of rotation of the disc of the host galaxy: the jet is compressing and accelerating the surrounding interstellar clouds. This behavior is therefore naturally accounted for by the jet/cloud interaction.

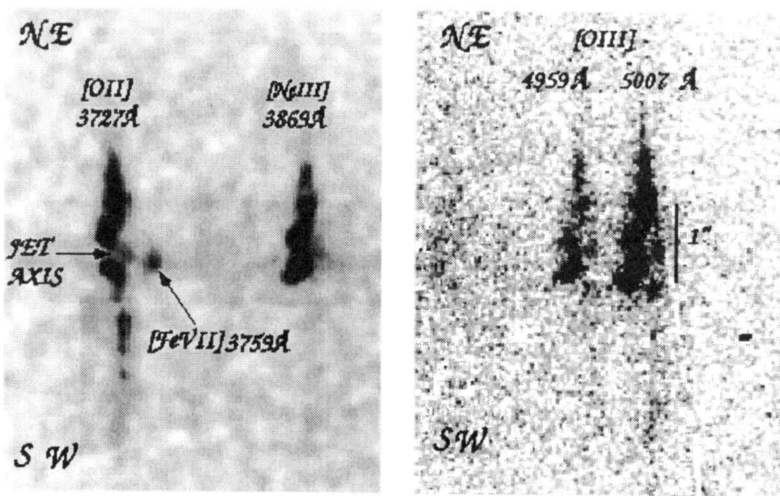

FIGURE 3: Grey-scale images of the emission line spectra of NGC 1068 obtained $\sim 1''$ North of the nucleus, with the slit oriented perpendicularly to the radio jet axis. The left–hand panel is centered around $[OII]\lambda 3727$ Å while the right hand panel is centered around $[OIII]\lambda\lambda 4959, 5007$ Å.

4. THE EXCITATION OF THE NLR

The HST images of NGC 1068 have also shown evidence for significant variations in the ionization structure within the emission line region which requires a fundamental change in the standard picture of the NLR (Capetti, Axon and Macchetto 1997). The material along the radio-jet, although denser than the surrounding gas, is in a much higher ionization state (as derived from the [O III] / (Hα + [N II]) ratio image, Fig. 4) than the surrounding regions of the NLR. We believe that this requires the presence of a source of ionizing radiation associated with the radio jet which locally dominates over the nuclear emission.

Axon et al. (1997a) showed that the radio-jets in the Seyfert 2 galaxies Mrk 348 and Mrk 3 are associated with an extended linear structure in UV and optical continua. They suggested that this extension is due to free-free emission produced by hot gas, shocked by the radio-jet which could provide a significant source of ionizing photons.

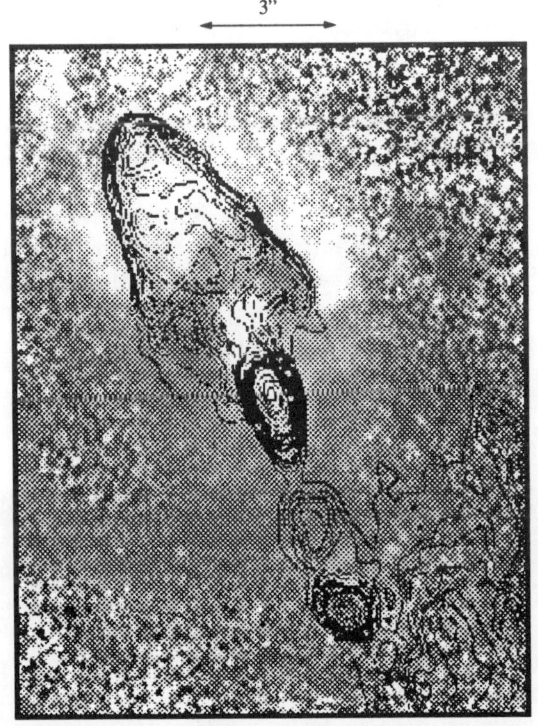

FIGURE 4: Radio contours of NGC 1068 superposed to the [O III] / (Hα + [N II]) emission-lines-ratio image. Note the region of high ionization (lighter color) co-spatial with the radio-jet. The outer boundary between the high and low (darker color) excitation regions corresponds to the transition between the jet-like and lobe-like morphology in the radio emission.

The observed ionization structure in NGC 1068 can be explained in a similar way provided that there are shocks created by the jet which are sufficiently fast to be auto-ionizing (Sutherland et al. 1993). In this case the ionization parameter can be elevated locally along the jet even though the density is higher than the surroundings.

This ionization structure is also seen in the spectra shown in Fig. 3. In the proximity of the radio jet the [OIII] is stronger than [OII] while outside the jet the excitation conditions are much lower with [OII] being substantially brighter than [OIII]. On the jet axis the coronal line [FeVII]λ3759Å is also seen. In the framework of pure nuclear photoionization we expect to observe coronal line only relatively close to the central source where the ionization parameter can be sufficiently high. The detection of this high excitation line at large distance from the nucleus confirms the presence of high temperature gas produced by the shocks driven by the jet, which produce a substantial flux of ionizing photons.

5. CONCLUSIONS

HST observations indicate that the origin, dynamics and ionization structure of the NLR of Seyfert galaxies are governed by the interaction of the interstellar clouds with the radio emitting outflows. The sweeping and compression produced by the shocks propagating in the interstellar medium enhances the density and determines the morphology of the emitting regions. The dynamics of the NLR clouds is set by the acceleration induced by the relativistic plasma which also provides the external pressure required to confine them. Finally, the gas heated by the shocks seems to be an important source of ionizing photons which is crucial in determining the ionization structure of the NLR.

It therefore appears that jet/cloud interaction in the nuclear region of Seyfert galaxies provides us with a scenario which naturally accounts for all the observed properties of their NLR. It might be speculated that a similar framework might be applied in general to the emission line regions of AGN and in particular to radio loud ones, which harbor much more energetic outflows than those observed in Seyfert galaxies and are expected to strongly interact with the surrounding medium.

These observations also provide a new tool for investigating the physical properties of jets based on the analysis of their effects on the surrounding medium, e.g. through numerical simulations of jet/cloud interaction effectively constrained by the observations of the line emitting gas associated with the regions of radio emission. This approach is discussed in detail by Rossi and Capetti (Chapter 13).

REFERENCES

Antonucci, R.R.J., & Miller, J.S., 1985, *ApJ*, **297**, 621.

Axon, D.J., Hough, J.H., Young, S.J., & Inglis M., 1994, in *Kinematics and Dynamics of diffuse astronomical media*, Kluwer Academic Press, ed. J.E. Dyson.

Axon, D.J., Capetti, A., Macchetto, F.D., & Sparks, W.B., 1997a, in preparation.

Axon, D.J., Macroni, A., Macchetto, F.D., Capetti, A., & Robinson, A. 1997b, in preparation.

Baldwin, J.A., Wilson, A.S. & Whittle, M., 1977, *ApJ*, **319**, 84.

Bower, G., Wilson A., Morse, J.A., Gelderman, R., Whittle, M., & Mulchaey, J., 1996, *ApJ*, **454**, 106.

Capetti, A., Macchetto, F.D., Axon, D.J., Sparks, W.B., & Boksenberg, A., 1995a, *ApJ*, **448**, 600.

Capetti, A., Axon, D.J., Kukula, M., Macchetto, F.D., Pedlar, A., Sparks, W.B., & Boksenberg, A., 1995b, *ApJ*, **454**, 85.

Capetti, A., Axon, D. J., Macchetto F. D., Sparks, W. B., & Boksenberg, A., 1996, *ApJ*, **469**, 554.

Capetti, A., Macchetto, F.D., & Lattanzi, M. G., 1997, *ApJ*, **476**, L67.

Capetti, A., Axon, D.J., & Macchetto, F.D., 1997, *ApJ*, in press.

Ferland, G. J., & Osterbrock, D. E., 1986, *ApJ*, **300**, 658.

Haniff, C. A., Wilson, A. S., & Ward, M. J., 1988, *ApJ*, **334**, 104.

Macchetto, F.D., Capetti, A., Sparks, W.B., Axon, D.J., & Boksenberg, A., 1994, *ApJ*, **435**, L15.

Meaburn, J., & Pedlar, A, 1986, *A&A*, **159**, 336.

Muxlow, T.W.B., Pedlar, A., Holloway, A.J., Gallimore, J.F., & Antonucci, R.R.J., 1996, *MNRAS*, **278**, 854.

Pedlar, A., Meaburn, J., Axon, D.J., Unger, S.W., Whittle, D.M., Meurs, E.J.A., Guerrine, N., & Ward, M.J., 1989, *MNRAS*, **238**, 863.

Sutherland, R.S., Bicknell, G.V., & Dopita, M.A., 1993, *ApJ*, **414**, 506.

Taylor, D., Dyson, J.E, & Axon, D.J., 1992, *MNRAS*, **255**, 351.

Tully, R.B., 1988, Nearby Galaxies Catalog, Cambridge University Press.

Unger, S.W., Pedlar, A., Axon, D.J., Whittle, M., Meurs, E.J.A., & Ward, M.J., 1987, *MNRAS*, **228**, 671.

Wilson, A.S., & Ulvestad, J.S., 1983, *ApJ*, **275**, 8.

Wilson, A.S., & Tsvetanov, Z.I., 1994, *AJ*, **107**, 1227.

CHAPTER 13

JET-CLOUD INTERACTIONS IN SEYFERT GALAXIES: NUMERICAL SIMULATIONS

P. ROSSI[1] AND A. CAPETTI[2]

[1] *Osservatorio Astronomico di Torino, Strada dell'Osservatorio 20, I-10025 Pino Torinese, Italy*

[2] *Scuola Internazionale di Studi Superiori Avanzati, Via Beirut 2-4, I-34014 Trieste, Italy*

We present preliminary results of numerical simulations of the interaction of a jet and a cloud in order to better understand the formation of the narrow line region in Seyfert Galaxies. We approach the problem by adopting a quite sofisticated hydrodynamical code and studying in detail the jet hydrodynamics, while we choose a more simplified treatment of radiative processes, in order to give a qualitatively good interpretation of the emission line processes.

1. INTRODUCTION

The recent discovery of radio jets in Seyfert galaxies and their importance in forming the narrow line region (Capetti, Chapter 12) motivates to perform numerical simulation of radiative jets interacting with the external medium. As in the case of stellar jet, most of the radiation is produceed as emission lines, while synchrotron radiation is negligible.

Many numerical studies have been devoted to the analysis of the propagation of a supersonic jet shot into an ambient medium. The primary interest of these studies was addressed to interpret the main features of the interaction between jet and enviroment visible in the powerful radio source maps. For this reason most of the numerical simulations only consider the adiabatic approximation since radiative losses (mainly synchrotron emission) are non crucial for this application. Radiative losses have been consider in fact only for stellar jets (Blondin, Fryxell and Königl 1990). The jet interaction with the external medium is clearly a very complicated physical problem which involves both a hydrodynamical study of the jet propagation as well as a detailed understanding of the microphysics of the induced shocks, and of the radiative processes. We intend approaching the problem studying in detail the jet hydrodynamics while adopting a simplified treatment of the radiative processes assuming the Raymond–Smith cooling function.

The consistency with the observed properties of the Narrow Line Region (NRL) is checked against their morphology, line-profiles and characterstic values of physical parameters of the emitting gas, such as density and temperature.

2. THE PHYSICAL PROBLEM AND THE SOLUTION METHOD

We study the evolution of a cylindrical fluid jet governed by the equations for mass and momentum conservation, and subject to radiative losses, whose specific form will be discussed later. The relevant equations are

$$\frac{\partial \rho}{\partial t} + \nabla \cdot (\rho \mathbf{v}) = 0 \,, \tag{1a}$$

$$\frac{\partial \mathbf{v}}{\partial t} + (\mathbf{v} \cdot \nabla)\mathbf{v} = -\nabla p/\rho \,, \tag{1b}$$

$$\frac{\partial E}{\partial t} + \nabla \cdot (E\mathbf{v}) = -p\nabla \cdot \mathbf{v} - \mathcal{L} \,, \tag{1c}$$

where the fluid variables p, ρ and $\mathbf{v}$ and E are, as customary, the pressure, density, velocity and thermal energy $(p/(\Gamma - 1))$, respectively; Γ is the ratio of the specific heats. $\mathcal{L}$ represents the energy loss term (energy lost per unit volume per unit time, Raymond and Smith 1977; Brinkmann et al. 1988).

The jet initially occupies a cylinder of length L and the initial flow structure has the following form:

$$
v_z(r) = \begin{cases} \dfrac{v_z(r=0)}{\cosh[(r)^m]} \,, & z \leq L \,, \\[2em] 0 & , \quad z > L \,. \end{cases}
$$

where m is a 'steepness' parameter for the shear layer separating the jet from the external medium. The choice of separating the jet's interior from the ambient medium with a smooth transition, instead of a sharp discontinuity, avoids numerical instabilities that can develop at the interface between the jet's proper and the exteriors, expecially at high Mach numbers.

The density radial dependence has the form:

$$
\frac{\rho(r)}{\rho(r=0)} = \nu - \frac{\nu - 1}{\cosh[(wr)^n]} \,.
$$

We have carried out our calculations setting $w = 0.75$, $m = 8$ and $n = 2m$; this implies a narrower and smoother radial extension of the 'density' jet with respect to the 'velocity' jet. The reason for this choice is to obtain a smooth radial profile of the momentum density ρv_z.

Integration is performed in cylindrical geometry and the domain of integration ($0 \leq z \leq D$, $0 \leq r \leq R$) is covered by a grid of 816×410 grid points. The axis of the beam is taken coincident with the bottom boundary of the domain ($r = 0$), where symmetric (for p, ρ and v_z) or antisymmetric (for v_r) boundary conditions are assumed. At the top boundary ($r = R$) and right boundary ($z = D$) we choose free outflow conditions, imposing for every variable Q null gradient ($dQ/d(r, z) = 0$). These free conditions do not completely avoid back-reflection phenomena from the outer boundaries. In order to limit this effect, the boundaries should be placed as far as possible from the region of the jet where the most interesting evolutionary effects presumably take place; for this purpose we employ a non-uniform grid both in the longitudinal (z) and the radial (r) directions. This non-uniform grid has high resolution in the region where the jet's head interacts with cloud. The numerical scheme adopted is of PPM (Piecewise Parabolic Method) type and is particularly well suited for studying highly supersonic flows with strong shocks (Woodward and Colella 1984, see also Bodo et al. 1994, 1995).

3. RESULTS

We report here some preliminary result of the numerical simulations performed as described above, but, before starting the description of these results, it is useful to remind the typical values of density, temperature and velocity for the emitting material in the NLR of Seyfert galaxies (see Chapter 12) deduced from observations. Typical densities are larger than 10^3 cm^{-3}, typical temperatures are $\sim 10^4 - 2 \times 10^4$ K, while the observed line widths correspond to velocities $\sim 300 - 1000$Km/s. These are the values of physical parameters that we have to match in our simulations. We considered, as a first attempt, the interaction of the jet with a uniform external medium ten times denser than the jet (jet density $= 0.1$ cm^{-3}; external density $= 1$ cm^{-3}), but we could not reach high enough compression factors to get density values comparable to the observed ones. We then considered an inhomogeneous external medium and thus the interaction of the jet with a cloud, which allows us to start with higher values of density. The parameters that characterize a simulation are therefore the jet Mach number M, the jet density n_j, the external density n_{ext}, the cloud density n_c, the jet temperature T_j and the external temperature T_{ext}. The cloud temperature is fixed by the pressure equilibrium condition between cloud and external medium, the jet, instead, is not in pressure equilibrium, but is overpressured with respect to the exterior. Here we present the results of a perticular simulation with the following parameters: $M = 500$, $n_j = 0.1$ cm^{-3}, $n_{ext} = 1$ cm^{-3}, $n_c = 100$ cm^{-3}, $T_j = 10^6$ K, $T_{ext} = 10^4$ K. For a discussion of the relative importance of the different parameters and of the dependence of the results on the choice of parameters see Rossi et al. (1997).

In Fig. 1 we show a snapshot of the density distribution at a time at which the jet has already bored its way through the cloud, the three small panels are enlargements of the region of interaction between jet and cloud showing respectively density, temperature and a quantity that should mimic the emissivity of the material given by the square of density for a temperature range $5 \times 10^3 - 2 \times 10^4$ K. We see that the emission is concentrated in a thin layer of compressed cloud material, whose width and mass grow in time as the shocked cloud material cools down. In Fig. 2 we have then represented the behavior of density, temperature and velocity along cuts through this layer. We can see that the material at $T \sim 10^4$ K has a density of about 1500 cm^{-3}, and we expect that the material at $t \sim 10^6$ K will soon cool down, increasing its density, since the cooling time from $T \sim 10^6$ K to $T \sim 10^4$ K is quite fast, as we will see below. The velocities of the material, in the interesting temperature range, are

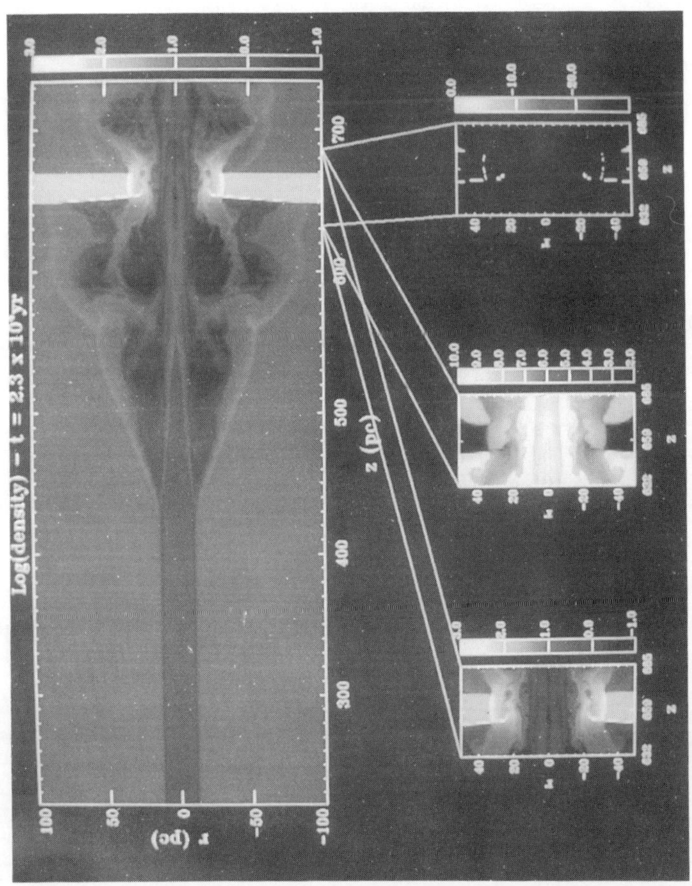

FIGURE 1. The top pannel show a snapshot of $Log(n)$ relatively to jet/cloud interaction, while in the three low panels are shown, from left to right, the zoom of $Log(n)$, the $Log(T)$ distribution and in the last one is a sort of *emission tracer* proportional to $Log(n^2)$ for a temperature range between 5×10^3 and 2×10^4 K (see Color Plate 2).

~ 200 km s^{-1}, giving a total line width of ~ 400 km s^{-1}, in the lower end of the observed range. Higher velocities can be seen for the higher temperature material.

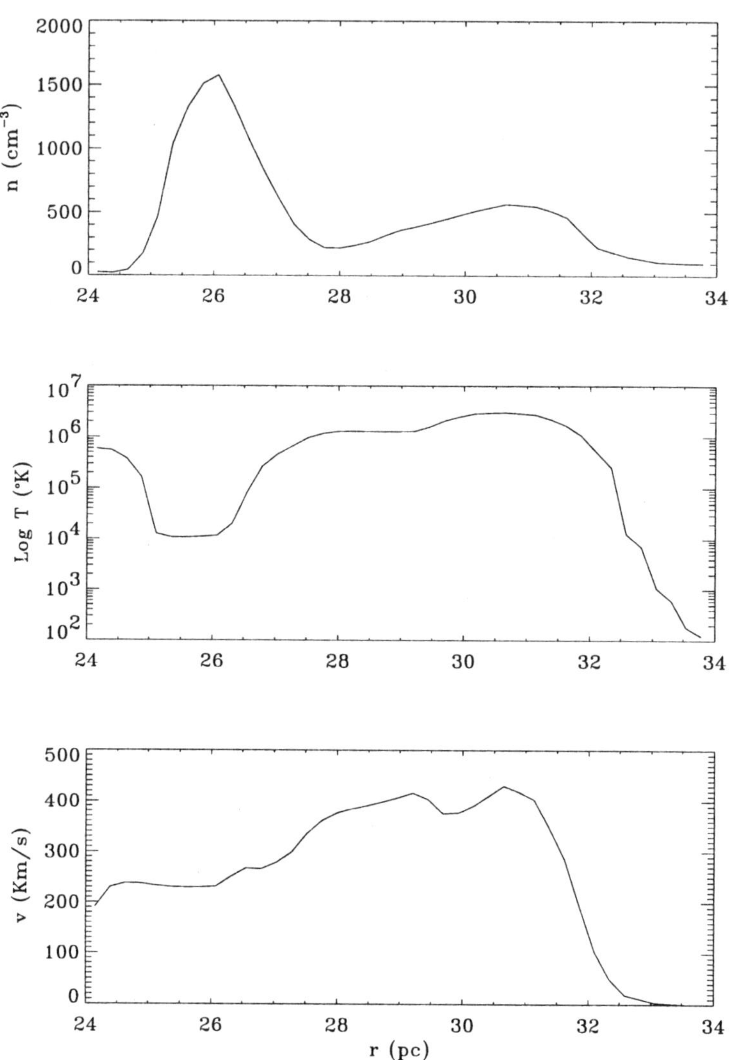

FIGURE 2. Plots of density (upper panel), temperature (middle panel) and velocity (lower panel) along a cut through the emitting layer. The position of the cut is at z=655 pc, the r coordinate on the abscissa represents the distance from the jet axis and the time is $t = 2.3 \times 10^4$ yr.

Figure 3 represents the temporal behavior of the total mass of the line emitting material (i.e. in the temperature range $5 \times 10^3 - 3 \times 10^4$ K), the different curves are for different lower density limits: we see that the growth of the emitting mass starts around $t \sim 10^4$ yrs, some time after the jet has begun to penetrate into the cloud (the jet first touches the cloud at $t \sim 4 \times 10^3$ yrs). This delay corresponds to the cooling time of the shocked material. We can compare Fig. 3 with Fig. 4, whose three panels represent respectively the temporal behavior of the maximum compression factor and of temperature and velocity at the point of maximum compression. We see that we have a first phase in which the maximum compression factor is lower than 10, up to $t \sim 10^4$ yrs, when, after this time, the compression factor grows beyond 10, we observe also a sensible growth of the emitting mass. In this phase, however, the densities of the emitting material are still quite low ($< 200 \mathrm{cm}^{-3}$ and the temperature at the point of maximum compression is still quite high ($\sim 10^6$K). At $t \sim 1.5 \times 10^4$ yrs we observe a fast drop in the temperature (the cooling time has become in fact very short) which is accompanied by another growth of the compression factor and of the emitting mass. In this stage we observe also a sensible growth of the emitting material at large densities (dashed-dotted curve in Fig. 3). The expansion velocities (lower panel in Fig. 4) are large in the initial phases of the evolution and then decay reaching an almost constant value around 200 Km/s.

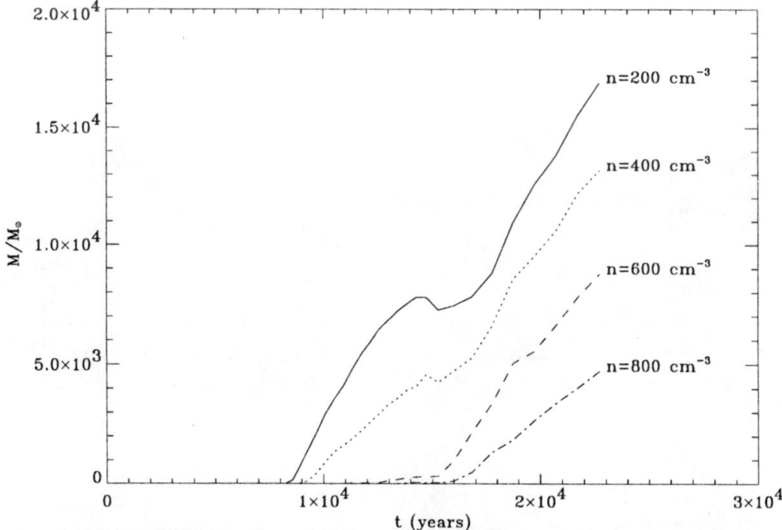

FIGURE 3. Temporal behavior of the emitting mass (in solar mass unit) at four different density limits.

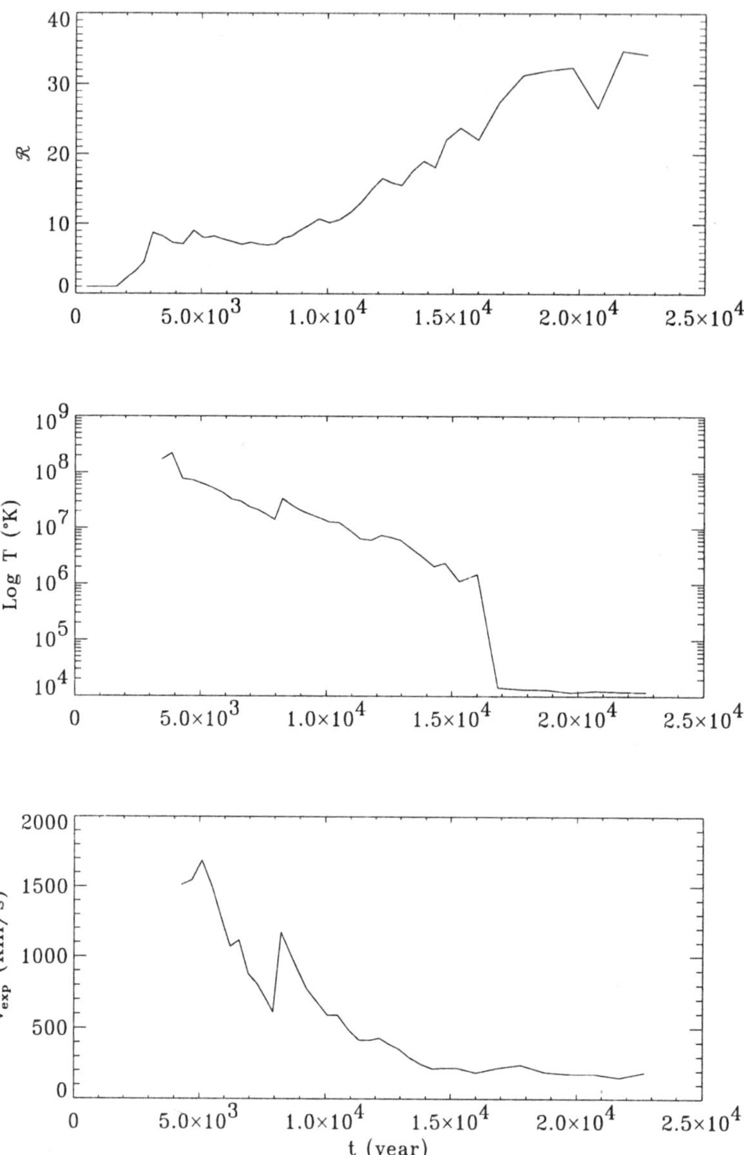

FIGURE 4. Plots of the temporal evolution of the maximum compression factor (upper panel), $\log(T)$ (middle panel) and expansion velocity (lower panel). Temperature and expansion velocity refer to the point at which the maximum compression factor is found.

4. SUMMARY

We have presented preliminary results of numerical simulations of the interaction between a radiative jet and a cloud for reproducing the physical conditions of the emitting material in the NLR of Seyfert galaxies. We found that an homogeneous medium can not give the values of density deduced from the observations, that is why we introduced a cloud in the ambient medium. In the interaction with this cloud we can in fact get high enough densities in the temperature range of the line emitting material; also the values of the expansion velocities and of the total emitting mass are in the range allowed by observations. Although we need more detailed analyses and a wide coverage of the parameter space, we can conclude that this model seems promising in explainig the origin and properties of the NLR in Seyfert galaxies.

REFERENCES

Blondin, J.M., Fryxell, B.A., & Königl, A., 1990, *Ap.J.*, **360**, 370.

Bodo, G., Massaglia, S., Ferrari, A., & Trussoni, E., 1994, *A&A*, **283**, 655.

Bodo, G., Massaglia, S., Rossi, P., Rosner, R., Malagoli, A., & Ferrari, A. 1995, *A&A*, **303**, 281.

Brinkmann, W., Fink, H.H., Massaglia, S., Bodo, G., & Ferrari, A., 1988, *A&A*, **196**, 313.

Colella, P., & Woodward, P.R., 1984, *J. Comp. Phys.* , **54**, 174.

Raymond, J.C., & Smith, B.R., 1977, *Ap.J.S*, **35**, 419.

Rossi, P., Capetti, A., Bodo, G., Massaglia, S., & Ferrari, A., 1997 in preparation.

CHAPTER 14

THE LONG TERM EVOLUTION OF POWERFUL RELATIVISTIC JETS

J. Mª MARTI[1], E. MÜLLER[2], J. Mª IBAÑEZ[1]

[1]*Departamento de Astronomía y Astrofísica*

Universidad de Valencia

E–46100 Burjassot (Valencia), Spain

[2]*Max–Planck–Institut für Astrophysik*

Karl–Schwarzschild–Str. 1

D–85740 Garching, Germany

We present results on the long term evolution of powerful extragalactic jets based on hydrodynamical relativistic simulations. Calculations have been performed with a high–resolution shock capturing code (Martí et al. 1995, 1997) that solves the equations of relativistic hydrodynamics in cylindrical coordinates. After a short initial transient phase in which the jet dynamics is governed by one–dimensional ram pressure arguments, the jet evolution is dominated by a strong deceleration phase during which the jet decelerates from speeds of about $0.26c$ to speeds smaller than $0.10c$ due to the degradation of the beam flow through internal shocks and the broadening of the beam cross section near the hot spot. The time evolution of jet/cavity overall quantities is studied and interpreted within a simple relativistic generalization of the Begelman and Cioffi's (1989) model for the evolution of overpressured cocoons surrounding powerful jets.

1. INTRODUCTION

Jets in powerful extragalactic radio sources (FR II radio galaxies and quasars) seem to have been feeding the lobes in these sources for periods as long as 10^7 years with kinetic powers in the range 10^{44}–10^{47} erg s^{-1} (Rawlings and Saunders 1991; Daly 1995). These powerful jets are supposed to be relativistic on parsec scales as indicated by a number of arguments (superluminal motion, one–sidedness, depolarization asymmetries, ...; see, e.g., Laing 1996). On larger scales, flux asymmetries between jets and counter–jets indicate that relativistic motion follows up to kpc scales, although with smaller values of the overall bulk speeds (Bridle et al. 1994), making it necessary to look for adequate models of flow deceleration between both scales. Finally, effective deceleration should occur at the terminal hot spots for which advance speeds in the range $0.01 - 0.1c$ are inferred (Liu, Pooley and Riley 1992; Daly 1995).

Whereas the numerical simulations of jets (performed extensively during the last fifteen years; see, e.g., Norman 1993, 1996) have proven to be very successful in the interpretation of the gross morphological and dynamical properties of radio jets, very few of these simulations have been followed long enough to account for the long term evolution of these sources. The simulation presented here represents a first step in a thorough study of the long–term evolution properties of powerful (relativistic) jets.

2. LONG TERM EVOLUTION OF POWERFUL RELATIVISTIC JETS

2.1. Hydrodynamic equations, numerical techniques and initial setup

Our simulation has been performed with a high–resolution shock capturing code (Martí et al. 1995, 1997) that solves the equations of relativistic hydrodynamics in cylindrical coordinates. In our model, the jet material is represented by an ideal gas of adiabatic exponent, γ. Interested readers are addressed to Martí et al. (1997) where the differential equations as well as their finite–difference form and an exhaustive testing of the hydrodynamic code can be found.

The code works in units in which the light speed, the jet radius at the injection point (R_b) and the density in the ambient medium (ρ_a) are all set to unity. The scaling to real sources is made by assigning values to R_b and ρ_a. Then the beam initial data are set by fixing the proper rest–mass density, ρ_b, the flow velocity v_b, the Mach number

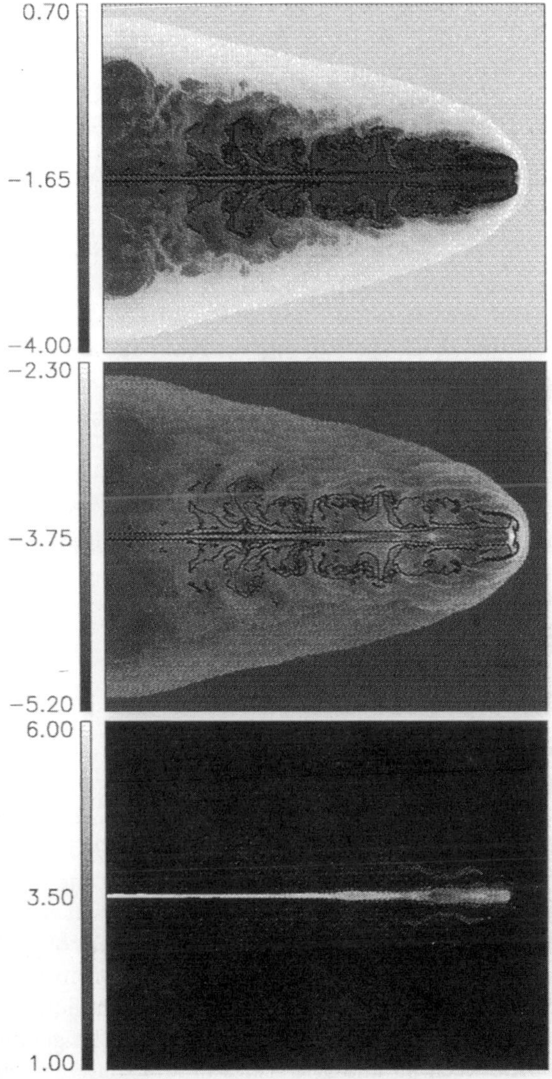

FIGURE 1. Logarithm of rest–mass density (top panel), logarithm of pressure (middle) and Lorentz factor (bottom panel) of the model discussed in the text at the end of the simulation. Black contours in the upper panels follow levels of constant (10%, 90%) beam particle fraction. They have been included to help recognizing the beam boundary and the interface between the lobe and the shocked external medium. Although the latter interface is defined somewhat arbitrarily, we find its evolution consistent with the expected lobe inflation occurring in the late phases of evolution.

M_b, and the constant adiabatic index γ. In the simulation presented here initial data are chosen to be $\rho_b = 3.4\,10^{-3}$, $v_b = 0.986$, $M_b = 9.0$ and $\gamma = 5/3$. Finally, the jet is assumed to propagate through a uniform, static atmosphere.

In the case of a highly–supersonic jet as the one presented here, the jet kinetic power P_b is given by

$$P_b = \pi R_b^2 \rho_b W_b (W_b - 1) v_b,$$

where $W_b = 1/\sqrt{1 - v_b^2}$ is the corresponding beam flow Lorentz factor, (in the present simulation, $W_b = 6.0$). Data are chosen to represent a powerful jet ($P_b = 10^{47}$ erg s^{-1}) when scaled according to $R_b = 0.35$ kpc and $\rho_a = 10^{-26}$ g cm^{-3}. From the jet setup a one–dimensional estimate of the hot spot (HS) propagation speed, v_{hs}^{1D}, can be derived (see Martí et al. 1994, 1997), which in the present model is $v_{hs}^{1D} = 0.26c$. The evolution has been followed up to $t = 2.7\,10^3$ in code units, $\approx 3\,10^6$ years, for our scaled model.

Our simulation has been performed with a resolution of 4 cells/R_b and the numerical grid spans a region of 85 R_b in radial direction and 220 R_b in axial direction. Phases I and II of the jet evolution (see below) have been repeated with 8 cells/R_b to check the sensitivity of our results on the grid resolution.

2.2. Overall morphology

Black and white coded snapshots of the proper rest–mass density, thermal pressure and flow Lorentz factor at the end of our simulation ($t = 2680 R_b/c$) are shown in Fig. 1. The overall morphology is typical of a light, highly–supersonic jet (Martí et al. 1995, 1997) and is characterized by an extended, overpressured cocoon, and a central jet with internal conical shocks. In our simulation (in which the resolution is too coarse to capture the process of entrainment of matter into the beam across its boundary) these shocks are responsible for the deceleration of the flow in the beam (see Fig. 1c). Beam flow speeds are largest near the injection point in agreement with the general trend of deceleration between pc and kpc scales in real sources. The cavity surrounding the jet and cocoon is also visible in the density and pressure panels as the region within the bow shock. As long as this region remains largely overpressured with respect to the ambient medium, its evolution is independent of the ambient pressure.

2.3. Long–term evolution

Figure 2 shows two panels with the HS propagation speed, v_{hs}, and the pressure, p_{hs}, as a function of time, respectively. The dashed line in the pressure plot represents the expected HS pressure, p_{hs}^e, assuming the HS to be a strong shock, $p_{hs}^e = \rho_a v_{hs}^2$. Both the expected and measured HS pressure nicely agree within a factor of 2 to 4. The values of v_{hs} and p_{hs} reached at the end of our simulation ($v_{hs} \approx 0.05c$; $p_{hs} \approx 0.002\rho_a c^2 \approx 2\,10^{-8}$ dyn cm^{-2}) are consistent with those inferred for powerful radiosources. Concerning the HS pressure, Carilli et al. (1996) report a value of $3\,10^{-9}$ dyn cm^{-2} for Cyg A's hot spot which coincides with our estimate within a factor of seven.

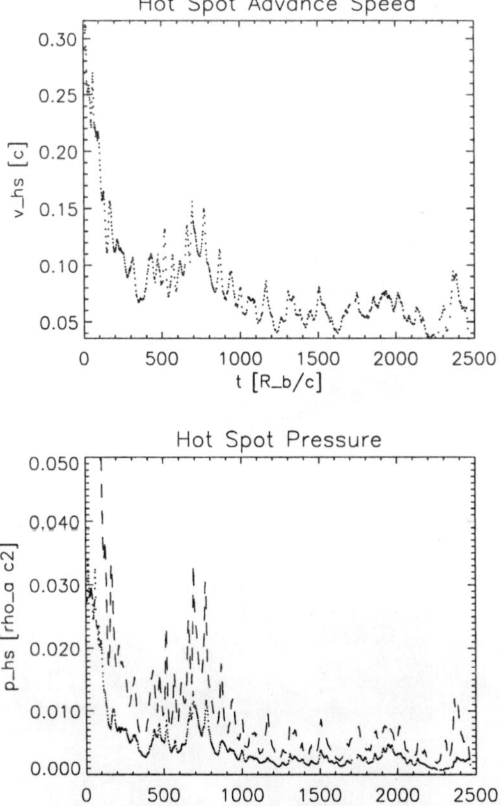

FIGURE 2. Hot spot advance speed (upper panel) and hot spot pressure (lower panel) for the powerful relativistic jet model discussed in the text. The dashed line in the pressure plot represents the estimated hot spot pressure assuming the hot spot to be a strong shock.

From the evolution of the HS advance speed (Fig. 2a), we conclude that the jet evolution is dominated by a strong deceleration phase occurring at early times. More precisely, we can distinguish four main evolutionary phases in the jet evolution. Phase I ($0 < t < 100$)

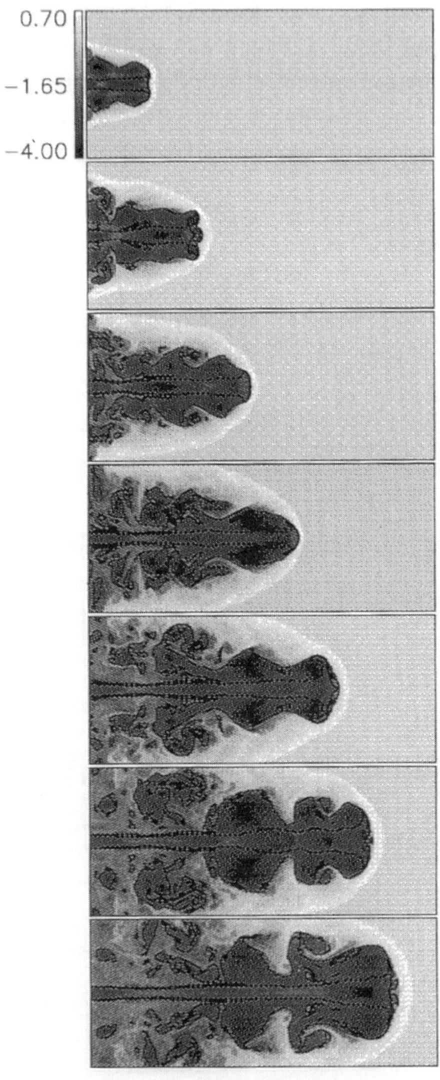

FIGURE 3. Evolution of the logarithm of proper rest–mass density during the period $0 < t < 350$ (Phases I and II). Jet deceleration and lobe inflation are clearly seen in the last four snapshots.

corresponds to a short initial transient phase in which multidimensional effects like internal beam shocks, variation of the beam cross section, etc... are still not important and the advance of the jet's working surface (the HS) is governed by one–dimensional ram pressure arguments. In Phase II ($100 < t < 350$) the jet decelerates from speeds of about $0.26c$ to speeds smaller than $0.10c$ due to the degradation of the beam flow through internal shocks and the broadening of the beam cross section near the HS. Figure 3 contains seven panels

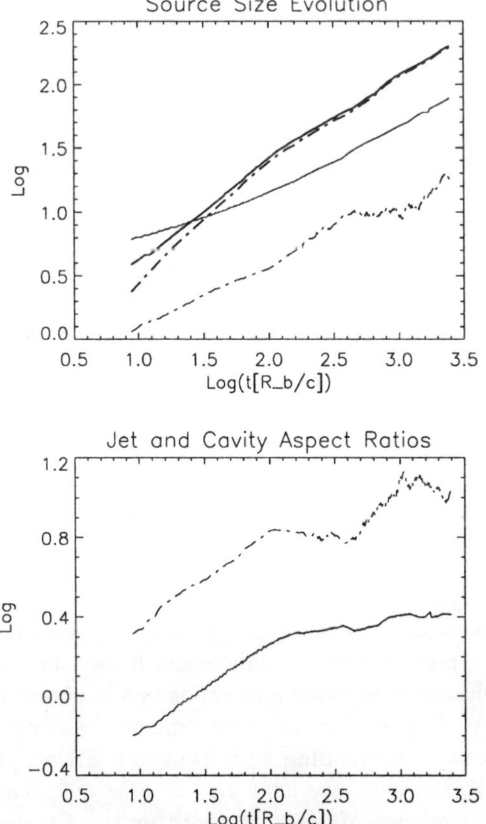

FIGURE 4. Top panel: cavity width (thinner filled line), cavity length (thicker filled line), lobe width (thinner dashed–dotted line) and jet length (thicker dashed–dotted line) as functions of time in a log–log plot. Bottom panel: cavity (filled line) and jet (dashed–dotted line) aspect ratios as a function of time.

with the proper rest–mass density covering the interval $0 < t < 350$ (Phases I and II). The strong deceleration is clearly observed in the last four panels in which big lobes of jet material start to inflate around the jet's head. The observations of many radio galaxies (e.g., Cyg A; Perley, Dreher and Cowan, 1984) and quasars (e.g., 3C154, 3C175, 3C334; see Leahy, Muxlow and Stephens, 1989) show the radio lobes in these sources in a process of inflation as if the jet would have decelerated and the flux of particles and energy into the lobe would have suddenly increased. A short acceleration phase (Phase III) follows from $350 < t < 750$. Finally, during Phase IV ($750 < t$), v_{hs} is small and evolves with a small power of time.

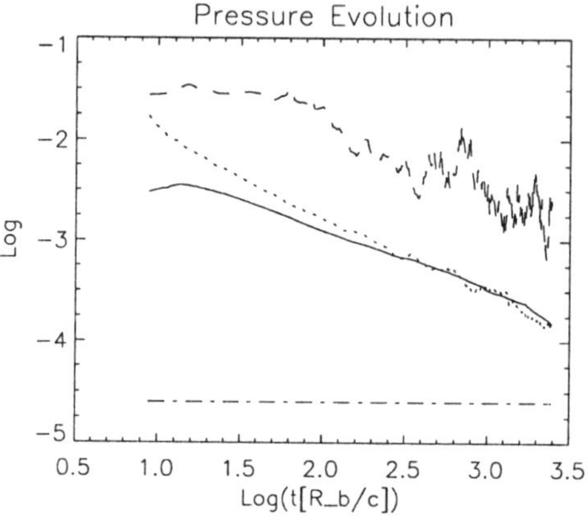

FIGURE 5. Mean cavity (filled line), lobe (dotted line) and hot spot (broken line) pressure as a function of time in a log–log scale. The (constant) ambient pressure is shown with a dashed–dotted line as reference.

Figure 4a displays the cavity's width and length (r_{cav}, l_{cav}, respectively), the lobe width (r_l) and the jet length (l_j) as functions of time in a log–log plot. Table 1 displays the linear slopes within each evolutionary phase corresponding to exponents in power–law fits. Phase I is characterized by an almost linear dependence of cavity and jet lengths with time corresponding to a constant HS propagation speed. In Phase IV, however, we have $l_{cav} \approx l_j \propto t^{0.6}$, leading to $v_{hs} \propto t^{-0.4}$. The dependence of the lobe width on the HS speed can be clearly deduced from Fig. 4a. Phase I, corresponding to an increase in lobe width proportional to $r_l \propto t^{0.4}$ is followed by a deceleration phase (Phase II) during which r_l increase faster due to the larger energy flux into the lobe through the jet terminal shock. During Phase III the input of energy into the lobe is nearly zero as expected in an acceleration phase. Finally, lobe inflation proceeds at an almost linear rate with time at late times (Phase IV). Cavity and jet aspect ratios (length/width) are shown in Fig. 4b as a function of time. There is a tendency for the cavity aspect ratio to become independent of time as

$p_{\rm hs}$ decreases and tends to the mean pressure in the cavity, $p_{\rm cav}$. The process of lobe inflation becomes apparent for the jet in the decrease of its aspect ratio with time ($\propto t^{-0.3}$). Finally, the time evolution of pressure is shown in Fig. 5. The HS pressure is almost constant ($\propto t^{-0.2}$) during the one–dimensional phase (Phase I) and decreases with time during the jet deceleration phases (Phases II and IV). In particular, the dependence of $p_{\rm hs}$ with time for $t > 750$ ($\propto t^{-0.8}$) is consistent with the one of $v_{\rm hs}$. The evolution of the pressure within the cavity can be estimated from the variation of thermal energy within the cavity, $E_{\rm cav}^{\rm th}$, and the variation of the cavity's volume as in the model of Begelman and Cioffi (1989). In their model,

$$p_{\rm cav} \approx E_{\rm cav}^{\rm th} l_{\rm j}^{-1} r_{\rm cav}^{-2},$$

with $E_{\rm cav}^{\rm th} \propto t$, $l_{\rm j} \propto t$ (assuming a constant HS advance speed) and $r_{\rm cav} \propto t^{0.5}$ (considering the bow shock that defines the cavity's volume in the limit of being a strong shock). Substituting these time dependencies in the previous equation one gets $p_{\rm cav} \propto t^{-1}$. However from our simulation we obtain $p_{\rm cav} \propto t^{-0.6}$ mainly due to the slower increase of the cavity volume as a consequence of the deceleration of the HS ($l_{\rm j} \propto t^{0.6}$ instead to $\propto t$; see also Cioffi and Blondin 1992).

TABLE 1: Power–law exponents

Phase	$l_{\rm cav}$	$r_{\rm cav}$	$l_{\rm j}$	$r_{\rm l}$	$A_{\rm cav}$	$A_{\rm j}$
I	0.81	0.35	0.93	0.43	0.46	0.47
II	0.60	0.48	0.62	0.71	0.13	-0.08
III	0.67	0.53	0.73	0.01	0.23	0.72
IV	0.59	0.58	0.60	0.88	0.01	-0.28

3. DISCUSSION AND CONCLUSIONS

We have shown results of the long term evolution of a relativistic, highly–supersonic, light jet that reproduces some properties observed in powerful extragalactic radio sources. A hot spot advance speed of about $0.05c$ is obtained after the deceleration phase occurring at early times in the source evolution. In our simulation, the beam flow is decelerated at internal shocks, which also cause the beam broadening that leads to the deceleration of the jet.

Our aim is to extend the study to a wide region in the ρ_b-v_b plane to constrain the values of these two fundamental parameters in powerful extragalactic radio sources. However, the validity of our present results can be limited by several factors: a) Two–dimensionality: three-dimensional effects can largely modify the stability of the jet and its propagation properties. b) Low resolution: the resolution used in our simulations is not enough to capture the process of mass entrainment which can alter the long term evolution. c) Short evolution: the time reached in our simulation is near the lower bound of the estimated ages for real sources; more realistic calculations should be performed covering a time period at least a factor of three longer than in our present simulation.

Acknowledgements

This research is supported by the Spanish DGICYT (Refs. PB94-0973, HA95-0190). J. Mª. Martí gratefully acknowledges a return grant from the HCM Program of the Commission of the European Communities (Contract ERBFMBICT950379).

REFERENCES

Begelman, M.C., & Cioffi, D.F., 1989, *Ap.J.*, **345**, L21.

Bridle, A.H., Hough, D.H., Lonsdale, C.J., Burns, J.O., & Laing, R.A., 1994, *AJ*, **108**, 766.

Carilli, C.L., Perley, R.A., Bartel, N., & Dreher, J.W., 1996, in *Cygnus–A Study of a Radio Galaxy*, C. L. Carilli, D. E. Harris. (eds.) Cambridge University Press, Cambridge, p. 76.

Cioffi, D.F. & Blondin, J.M., 1992, *Ap.J.*, **392**, 458.

Daly, R.A., 1995, *Ap.J.*, **454**, 580.

Laing, R.A., 1996, in *Energy Transport in Radio Galaxies & Quasars*, P. E. Hardee, A. H. Bridle, J. A. Zensus. (eds.), *ASP Conference Series*, Vol. **100**, p. 241.

Leahy, J.P., Muxlow, T.W.B., & Stephens, P.W., 1989, *MNRAS*, **239**, 401.

Liu, R., Pooley, G., & Riley, J.M., 1992, *MNRAS*, **257**, 545.

Martí, J.Mª., Müller, E., Font, J.A., & Ibáñez, J.Mª, 1995, *Ap.J.*, **448**, L105.

Martí, J.Mª., Müller, E., Font, J.A., Ibáñez, J.Mª & Marquina, A., 1997, *Ap.J.*, **479**, 151.

Martí, J.Mª., Müller, E., & Ibáñez, J.Mª, 1994, *A&A*, **281**, L9.

Norman, M.L., 1993, in Astrophysical Jets, D. Burgarella, M. Livio, C. O'Dea. (eds.), Cambridge University Press, Cambridge, p. 211.
Norman, M.L., 1996, in *Energy Transport in Radio Galaxies and Quasars*, P. E. Hardee, A. H. Bridle, J. A. Zensus. (eds.), it ASP Conference Series, **100**, p. 405.
Perley, R.A., Dreher, J.W., & Cowan, J., 1984, *Ap.J.*, **285**, L35.
Rawlings, S., & Saunders, R., 1991, *Nature*, **349**, 138.

CHAPTER 15

NONLINEAR INSTABILITIES IN SUPERSONIC JETS

G. BODO

Osservatorio Astronomico di Torino, Pino Torinese, Italy

I compare the results of two–dimensional and three–dimensional numerical simulation of jet instabilities showing that in three–dimensions the instability grows much faster and in a more disruptive way. This can be understood as a consequence of two main differences between the two cases: i) the linear growth of non–axisymmetric modes, which can not be present in the two–dimensional case; ii) the nonlinear cascade of energy towards small scales, which is known to be present only in the three–dimensional case. These results propose again the challenge of explaining the stability of the observed real objects.

1. INTRODUCTION

When the jet model for extragalactic radio sources was first proposed, one of the main concern was the stability of such collimated flows. In fact, hydrodynamical instabilities driven by the velocity shear (Kelvin-Helmholtz instabilities) seriously threaten the ability of the flow of transferring energy from the galactic nucleus to the extended lobes. After more than twenty years of research in this field, we are in the position of asking more detailed questions, but the problem of the effects of instabilities on jets is still open. This is largely due to the fact that, apart from the very first beginning of the instability evolution, when perturbations are small and the equations can be linearized, we confront with a higly nonlinear problem and, in order to get some result, we must mainly resort to numerical simulations.

Initially, the main focus was on the ability of jets to survive the instability over large distances, without losing most of their energy. However it was soon realized that some of the jet kinetic energy had to be dissipated, in order to get the radiation we see, and intabilities could provide the most natural way to perform this job. In addition, several different morphological features, such as knots or wiggles, which appear to be common in all the different classes of astrophysical jets, have been, in many instances, interpreted as the outcome of the instability growth. Not all the jets require the same degree of stability: jets in FRII radio sources appear to be very higly collimated, while the properties of FRI radio sources have been often interpreted as due to turbulent flows (Bicknell 1984). In the first case we need highly stable jets, while, in the second case, the instability might drive the transition to turbulence. In summary, the questions we are now facing have become more complicated and subtle: instability is required at some degree and some jets have to be more stable than others.

As I said before, the main approach for studying the instability evolution beyond the linear phase has been the use of numerical simulations, which are however limited by supercomputer power. Only recently three–dimensional simulations with sufficient resolution have become possible (Hardee and Clarke 1992, Hardee, Clarke and Howell 1995, Basset and Woodward 1995, Bodo et al. 1997) and most of the available results refer to the two–dimensional case (Norman and Hardee 1988, Hardee and Norman 1989, Hardee et al. 1992, Bodo et al. 1994, Bodo et al. 1995). However the evolution is inherently three-dimensional and the obvious question that I want to address in this Chapter is to what extent two–dimensional results can give an insight to the actual three–dimensional evolution.

2. LINEAR ANALYSIS

The linear stability analysis of jets has been the subject of a long series of studies, which have mainly considered compressible, supersonic flows, in cylindrical, slab or conical geometry, introducing several different physical ingredients, such as magnetic field, rotation, relativistic motion, anisotropic pressure, radiation, etc... (for a comprehensive review see Birkinshaw 1991). Some of these ingredients can have a stabilizing effect: magnetic field is the most important, however its stabilizing effect depends in a critical way on its structure. It seems therefore difficult to find a jet configuration which could be absolutely stable. The growth times of the instability are

quite fast, for the fastest growing modes they are comparable to the
jet flow time. On the basis of linear analysis alone, one would therefore conclude that the jet could be very soon destroyed. However,
the actual evolution, in which saturation effects could act efficiently,
can be determined only through the solution of the fully non–linear
problem.

One of the most important results of linear analysis is the clas-
sification of unstable modes, for compressible flows, in *ordinary* and
reflected modes. The first is a *surface mode*, in the sense that perturbations are maximal at the jet surface and rapidly decay away from
it; the others are present only in the compressible case, dominate the
instability behavior in the supersonic regime and are *body modes*, in
the sense that perturbations are not confined to the jet surface but
are present in the whole jet body.

3. NUMERICAL SIMULATIONS

Most of the numerical simulations that have been performed consider
the two–dimensional case. In two–dimensions one can consider either
cylindrical or planar geometry. In the first case it is only possible to
follow the evolution of axisymmetric perturbations (*pinching* modes);
in the second case, instead, we have two kind of modes: *symmetric*
and *antisymmetric* modes.

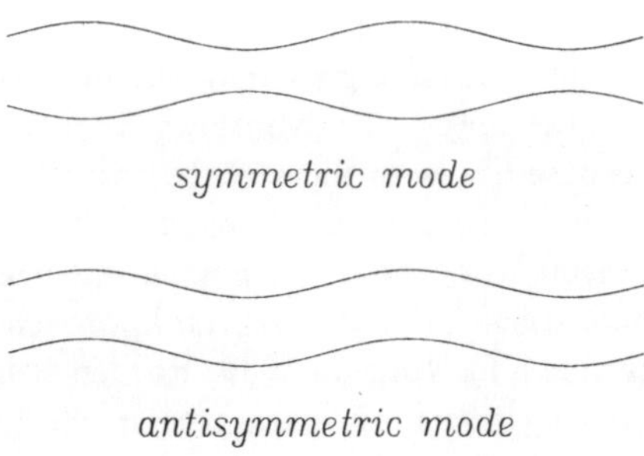

FIGURE 1. Deformation induced to the surface of a slab jet by symmetric
and antisymmetric perturbations.

In Fig. 1, I represent the deformation induced at the jet surface by the two kind of modes: the *symmetric* mode can be considered as the analog of the *pinching* mode of a cylinder, while the *antisymmetric* mode induces a shift to the jet body in a way which is analog to the shift induced by the *helical* mode of a cylinder. Linear stability analyzes have shown that the growth rates of the *symmetric* mode of the slab jet and of the *pinching* mode of the cylindrical jet are almost equal. Similarly, also the growth rates of the *antisymmetric* and *helical* modes are very similar. Based on those considerations, many authors have studied the behavior of a slab jet in two–dimensions, with antisymmetric perturbations, in order to get some insight about the evolution of the actual three–dimensional jet. In the following, I will first show the result of two–dimensional slab jet simulations (Bodo et al. 1995) and I will then compare them with three–dimensional results (Bodo et al. 1997). Both kind of simulations have been performed using a PPM code (Colella and Woodward 1984).

2.1. Two–dimensional slab

The initial configuration is a planar slab jet, separated by a smooth transition from the external medium at rest. The boundary conditions are periodic in the flow direction and free in the transverse direction. To this equilibrium structure we impose a superposition of periodic tranverse velocity disturbances (for details see Bodo et al. 1995). The choice of periodic boundary conditions in the flow direction implies that we follow the *temporal* evolution of the perturbation, an alternative choice would be that of continuously disturbing the left boundary and following the *spatial* growth of the perturbation. Both approaches have advantages and disadvantages, the choice of a *temporal* approach in this case has been dictated by the fact that we want to follow the long term jet evolution. Using a *spatial* approach one is limited by the crossing time of the perturbation over the length of the grid. For simulations that use a *spatial* approach see Norman and Hardee 1988, Hardee and Norman 1989, Hardee et al. 1992. The system is characterised by two main parameters: the Mach number M of the flow (with respect to the internal sound speed) and the density ratio ν between external medium and jet and we have widely explored the resulting parameter plane.

The jet evolution can be divided in four distinct phases, independently from the parameters. The particular choice of the parameters changes, instead, the relative length and importance of these different phases whose main characteristics are the following:

1) *Linear phase:* In this stage the modes excited by the initial per-

turbation grow in agreement with the predictions of linear theory. No interaction still occur between jet and external medium and the end of this phase is marked by the formation of internal shocks.

2) *Acoustic phase:* The growing jet deformation drives shocks in the external medium. These shocks can transfer energy and momentum from the jet to the external medium and the jet begins to slow down.

3) *Mixing phase:* As a consequence of the shock evolution, mixing between jet and ambient medium begins. The interaction with the ambient medium is then stronger and more momentum is transferred to it.

4) *Stationary phase:* In this final phase the system reaches a quasi steady state, whose characteristics depend on the choice of the physical parameters M and ν.

The main differences between jets with different Mach numbers and density ratio are that the mixing phase occurs earlier for high density and low Mach number jets, but it is more disruptive in the case of light jets. In the final stage, high density jets still keep a coherent structure and their average longitudinal velocity is not greatly decreased; low density jets, instead, do not exist any more as well ordered flows and maximal velocities are subsonic.

2.2 Three–dimensions vs. Two–dimensions

We can expect several differences between the three–dimensional and the two–dimensional evolution. First of all, in three-dimensions we have all the series of high order non–axisymmetric modes, which can not be followed in two–dimensions (apart from the *helical* mode, as already discussed). Linear stability analysis predict in many cases a growth rate for these modes which is higher than that of the *pinching* or of the *helical* mode. Furthermore it is known that properties of turbulence are different in three and two-dimensions: while, in the first case, energy cascades towards the small scales, the opposite occurs in the second case. This can change the behavior of jets in the mixing phase: mixing can be enhanced by the predominance of small scale structures. Finally, we have a different scaling law for the volumes: while, in the two–dimensional slab, volumes scale as the the first power of the radial extent, in three-dimensions volumes scale as the second power. Therefore, for a similar increase in radial extent, we must take into account that the entrained mass is much larger in three than in two–dimensions.

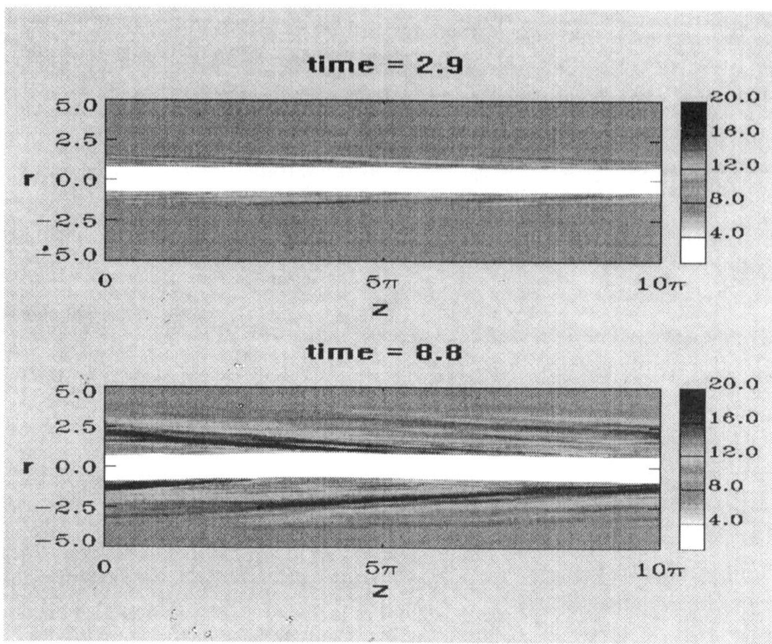

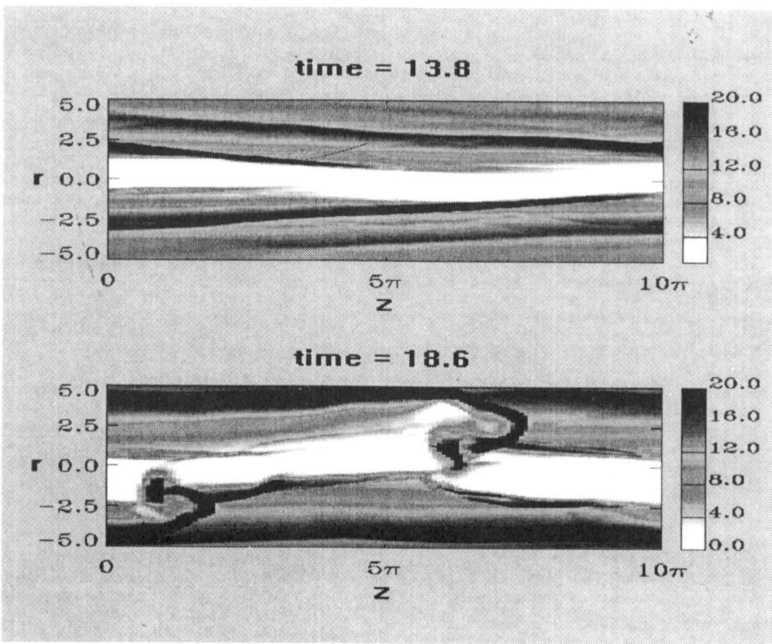

FIGURE 2. Gray scale images of the density distributions, at four different times, for a $M = 10$, $\nu = 10$ two-dimensional slab jet.

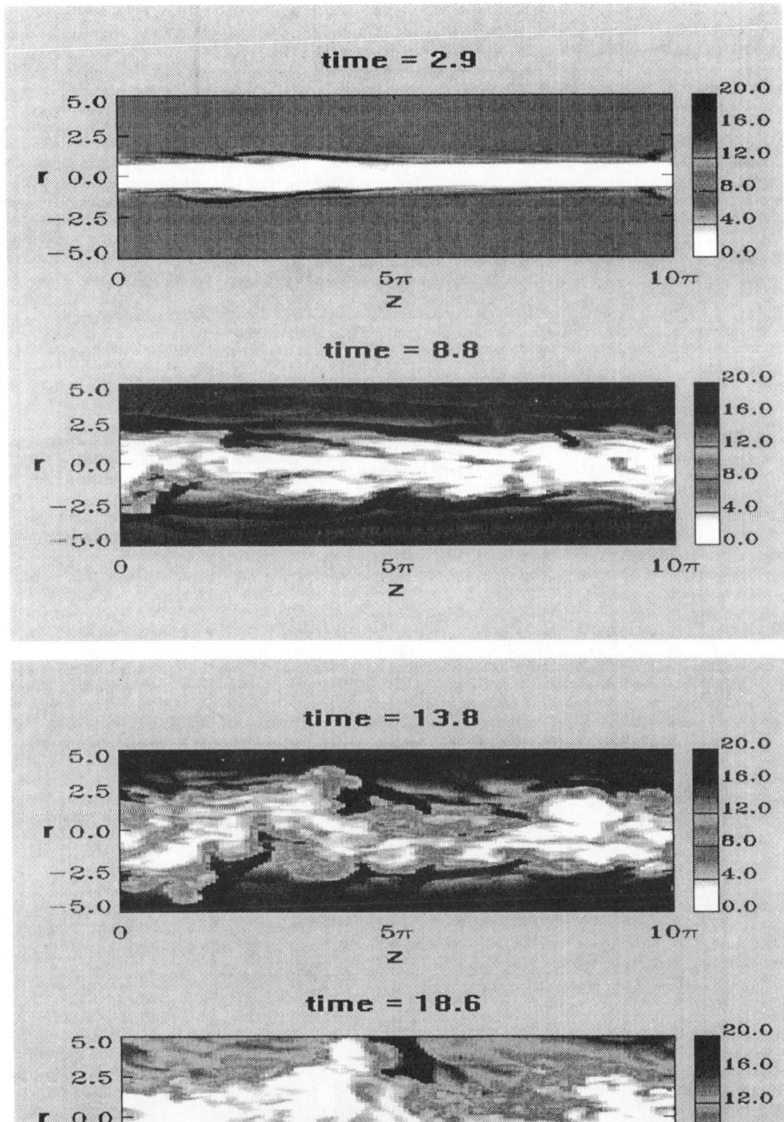

FIGURE 3. Gray scale images of the density distributions in a two-dimensional cut in the xy plane, at four different times, for a $M = 10$, $\nu = 10$ three-dimensional jet.

In Fig. 2, I show grey scale images of the density distribution, at four different times, for the case $M = 10$, $\nu = 10$ (light jet) for the two–dimensional slab jet case. These distributions can be compared with the distributions obtained in the three–dimensional case, at the same times, presented in Fig. 3. The much faster evolution in the three-dimensional case is immediately evident. Mixing begins to occur very early in the evolution, due to the fast development of small scale structures. In this case it appears that these structures are formed as a consequence of the growth of high order non–axisymmetric modes, which, in fact, have a higher linear growth rate (Bodo et al. 1997). The faster evolution in three–dimensions can be noticed also by looking at the behavior of the jet momentum as a function of time, which is represented in Fig. 4, where one can see the rapid decrease in the three-dimensional case (solid line) as compared to the two–dimensional case (dashed line).

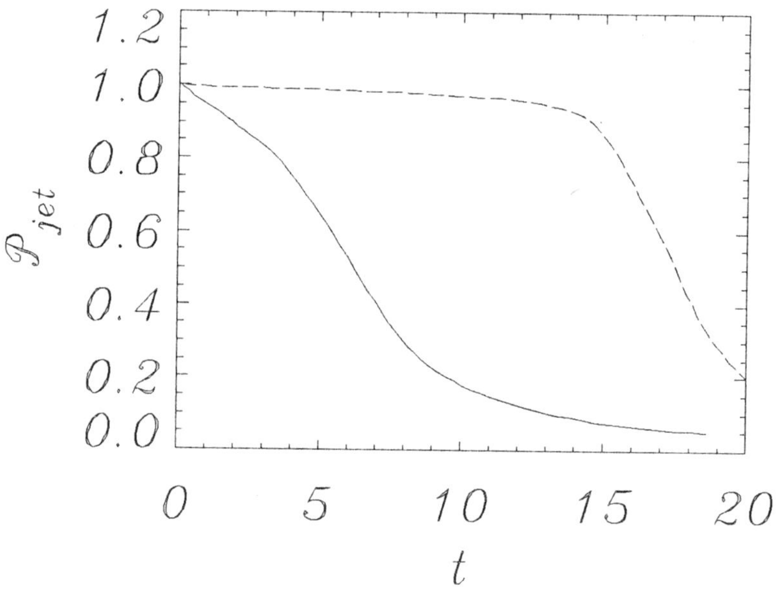

FIGURE 4. Plot of jet momentum as a function of time for the case $M = 10$, $\nu = 10$. The solid curve refer to the three–dimensional case while the dashed curve refers to the two–dimensional slab jet case.

The four phases of jet evolution, which were very well distinct in the two–dimensional evolution become intermixed in three–dimensions. More precisely, as I already noted, mixing starts very early, due to the fast growth of small scale structures, and the processes of momentum loss through shock waves driven in the external medium, characteristic of the *acoustic phase*, and through material mixing, characteristic of the *mixing phase*, occur simultaneuously; therefore, these two phases are not clearly distinguishable as before.

The case of a jet denser than the surrounding medium present a similar evolutionary pattern; the main difference is that the growth of small scale structures and the consequent mixing process occurs somewhat later and the first phases of evolution are more similar to the two–dimensional case. The final state presents, instead, a difference in the percentage of jet momentum lost to the ambient medium: while in the two–dimensional case the jet loses about 40% of its initial momentum, in the three–dimensional case this percentage grows up to almost 80%. This difference can be interpreted as a consequence of the different volume scaling discussed at the beginning of this Section: the mass entrained is larger in three-dimensions and this means a larger loss of momentum.

Finally, one can remark, from Figs. 2 and 3, that, at late times, the large scale three–dimensional jet morphology, which is mainly attributable to the evolution of the *helical* mode, still presents similarities with the two–dimensional slab. This suggests that the evolution of this mode can be indeed captured by two–dimensional calculations, but the global jet structure in three–dimensions is determined in large fraction by the evolution of small scale structures which are not present in those calculations.

4. CONCLUSIONS

Three–dimensional simulations of the non linear evolution of Kelvin–Helmholtz instabilities in supersonic, hydrodynamic jets show a very fast growth of unstable modes, which, very soon, induce strong material mixing and momentum transfer between jet and external medium. The evolutionary times, of the order of a few sound crossing times over the jet radius, are much faster than in the corresponding two–dimensional slab jet cases. One one hand, it is clear that the increase in dimensionality induces new physical effects that change in a substantial way the jet evolution: one has to be, therefore, very careful in drawing strong conclusions from two–dimensional results. On the other hand, these three–dimensional results raise again the question of how astrophysical jet manage to maintain the strong stability implied by observational data. I have considered here only hydrodynamic, adiabatic jets and, obviously, one could try to introduce many other physical effects that could have a stabilizing effect. Two possibilities are magnetic field and radiation. However, MHD numerical codes are not as robust as hydrodynamic codes and only very recently some results have been obtained in three–dimensions (Hardee, Stone and Rosen 1997) and more work is needed in this direction. As for radiation, two–dimensional results seem to imply a strong stabilization effect (see Chapter 20 and Rossi et al. 1997), but three–dimensional simulations are obviously needed.

Acknowledgements

The three–dimensional calculations have been performed on the Cray T3D of Cineca.

REFERENCES

Bassett, G.M., & Woodward, P.R., 1995, *Ap.J.*, **441**, 582.

Bicknell, G.V., 1984, *Ap.J.*, **286**, 68.

Birkinshaw, M.: 1991, *Beams and Jets in Astrophysics*, P.A. Hughes edt., Cambridge Univ. Press, Chap. 6.

Bodo, G., Massaglia, S., Ferrari, A., & Trussoni, E., 1994, *A&A*, **283**, 655.

Bodo, G., Massaglia, S., Rossi, P., Rosner, R., Malagoli, A., & Ferrari, A. 1995, *A&A*, **303**, 281.

Bodo, G., Rossi, P., Massaglia, S., Rosner, R., Malagoli, A., & Ferrari, A. 1997, *A&A*, submitted.

Colella, P., & Woodward, P.R., 1984, *J. Comp. Phys.*, **54**, 174.

Hardee, P.E., & Norman, M.L., 1989, *Ap.J.*, **342**, 680.

Hardee P.E., Cooper, M.A., Norman, M.L., & Stone, J.M., 1992, *Ap.J.*, **399**, 478.

Hardee, P.E., & Clarke, D.A., 1992, *Ap.J.*, **400**, L9.

Hardee, P.E., Clarke, D.A., & Howell, D.A. 1995, *Ap.J.*, **441**, 644.

Hardee, P.E., Stone, J., & Rosen, A., 1997, in *Low-Mass Star Formation from Infall to Outflow*, poster proceedings of the IAU Symp. n. 182, F. Malbet & A. Castets eds., p.132.

Norman, M.L., & Hardee P.E., 1988, *Ap.J.*, **334**, 80.

Rossi P., Bodo G., Massaglia S., Ferrari A., 1997, *A&A*, **321**, 672.

JETS IN OUR GALAXY

CHAPTER 16

JETS: A STAR FORMATION PERSPECTIVE

T.P. RAY

Dublin Institute for Advanced Studies, 5 Merrion Square, Dublin 2, Ireland

Current observations suggest as much as 10% of the matter that is accreted through discs around young stars ultimately gets ejected. The expulsion of this material is seen in a variety of ways including high velocity molecular (CO) outflows, shocked H_2 emission, and Herbig-Haro (HH) type jets. In this paper, I will briefly review our current knowledge of HH jets and show how a 'unified' model is now emerging in which HH jets are thought to be ultimately responsible for driving all of the other outflow components. Perhaps the biggest remaining challenge in this field is to understand the origin of the jets themselves. Some recent observations that give us clues as to how they are formed will be discussed.

1. INTRODUCTION

When I attended the first Turin Workshop in the early 1980s, the subject of jets from young stars had not been born. Then, and perhaps even to this day, the main emphasis was on understanding their much larger and more energetic counterparts i.e. jets from active galactic nuclei (AGN). While there are major differences between the two groups of jets, for example AGN jets are effectively adiabatic although this is clearly not the case for jets from young stars, there are some similarities. Both AGN and Young Stellar Object (YSO) jets often have similar morphologies with quasi-periodically spaced knots and 'wiggling' motion. Thus jets from young stars are not only interesting in themselves but, I think, can give us valuable insights into their bigger brethren. Moreover obtaining data on YSO jets is relatively easy (and for this reason I think we understand them better) because they are emission line objects and so we can determine their densities, temperatures, radial velocities, etc., spectroscopically. In comparison similar data on extragalactic jets tends to be much fuzzier. YSO jets also have the advantage that one can obtain proper motion data for various morphological features within a matter of months using the Hubble Space Telescope (HST), and usually within a few years from the ground. This information can then be combined with spectroscopic data to build up a rather thorough picture of YSO jet kinematics.

Historically the finding that YSOs have outflows (ironically proving inflow is present has turned out to be much more difficult!) can be traced back to the independent discovery by George Herbig and Guillermo Haro (Herbig 1951; Haro 1952) of faint emission line nebula in a number of star forming regions. In a seminal paper Osterbrock (1958) suggested that Herbig-Haro (HH) objects, as they became to be known, might be caused by energetic outflows from young stars. Almost twenty years, however, were to pass before Schwartz (1975) noticed the similarity of the optical spectra of HH objects with those of supernova remnants (SNRs) leading him to propose that, as in the case of SNRs, the emission arose from radiative shocks. At the same time there was already mounting evidence (e.g., Kuhi, 1964) for mass loss amongst YSOs and this suggested to Schwartz (1975) that HH shocks were ultimately driven by a stellar wind. Later Schwartz (1978) proposed that HH objects might arise as a result of shocks caused by 'blockages' in the flow, cloudlets that got in the way of a supersonic wind from a young star. Norman and Silk (1979) put forward an alternative scenario when they suggested that HH objects were shock waves in the ambient medium caused by YSO 'bullets'.

The discovery that many HH objects either mark the region where a jet 'terminates' (the reason from the quotation marks will be made clear later), or are actual parts of a jet, came with the discovery of YSO jets in the early 1980s. Prior to that a number of observations were made of YSO jets although they were not identified as such. For example Dopita, Schwartz and Evans (1982) found that the HH 46/47 flow, located in the Gum Nebula, was bipolar with red-shifted and blue-shifted HH objects being ejected on opposite sides of an embedded T Tauri star. Moreover they commented that the remarkable linear alignment of the HH objects in this system suggested 'a high degree of collimation in the bipolar flow'. They did not seem to realize however that the faint 'nebular bridge' (HH 47B) linking HH 47A to HH 46, and discovered several years before by Schwartz (1977), was in fact the actual northern jet from this embedded star. The first HH flows to be recognized as jets were from HL Tau, HH 30 and DG Tau B (Mundt and Fried 1983). Many more were then found (Mundt et al. 1984; Strom et al. 1985; Reipurth et al. 1986; Ray 1987) and today a large number of collimated HH flows are known (Reipurth 1994).

2. A TYPICAL EXAMPLE

Although it is sometimes dangerous in astronomy to choose an archetype, as they are often subsequently found to be pathological in some way (!), we will nevertheless take as our example the HH 34 flow since it illustrates many of the properties of YSO jets. Listed by Herbig (1974) as an 'irregular nebulous spot about 15″ across', it was independently discovered by Haro in 1959 [Clearly the pressure to publish was not as great then as it is now!] Despite being one of the brightest HH objects known, it really only attracted attention with the discovery of a very fine jet pointing towards it (Mundt 1986; Reipurth et al. 1986; Bührke, Mundt and Ray 1988) whereupon it was realized that HH 34 is in fact the bow shock of the HH 34 jet as it ploughs its way though its surroundings (see Fig. 1). Moreover a counter-bow shock (HH 34 N) was also discovered to the north (Bührke *et al* 1988) at approximately the same distance as HH 34 is from the jet source. More recently Bally and Devine (1994) have shown that HH 34 S (as the originally discovered HH object is now known), HH34 N and the HH 34 jet are part of a much larger outflow system stretching almost 0.5° in projection on the sky. The other HH objects in the flow include HH 172, HH 86, HH 87, and HH 88 to the south and HH 126, HH 85, and HH 33/40 to the north. Radial velocity measurements have shown that HH 34 S and the jet are blue-shifted while the northern counter-flow is red-shifted. The typical radial velocity of the jet is around -100 km s^{-1} (Bührke et al. 1988) although there is a slower component present along some parts of the flow. Similar radial velocities are found for the HH 34 S bow shock suggesting that the jet is meeting little resistance from the material ahead of it and that the reverse shock (i.e. the Mach disc) must effectively be of low velocity. The emission line profiles observed in the region of the HH 34 S bow have precisely the double component structure one expects on the basis of theory (Hartigan et al. 1987).

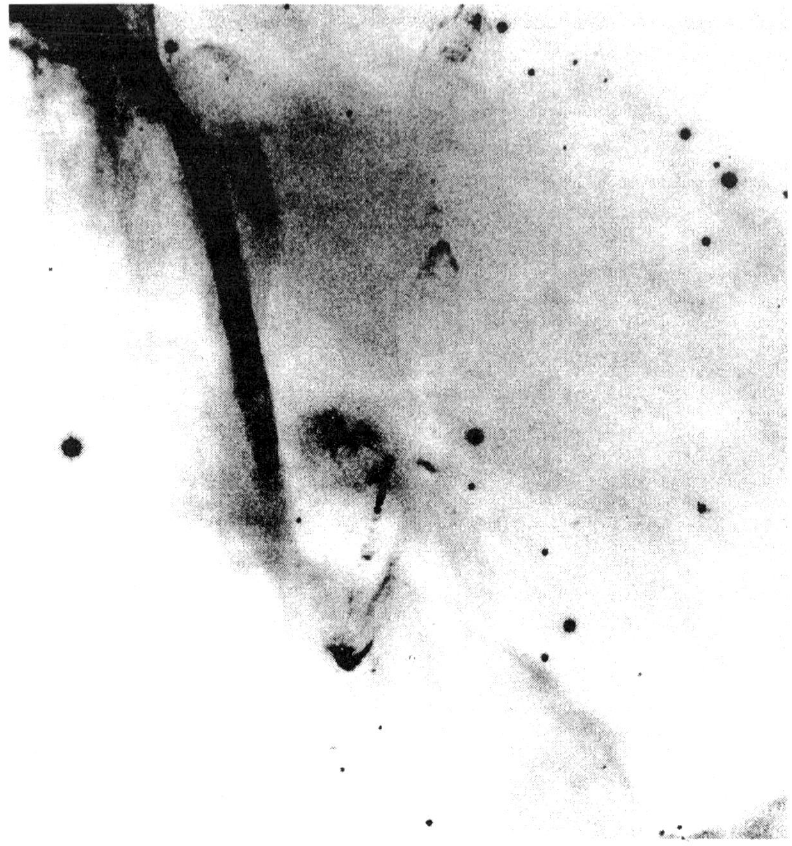

FIGURE 1. A narrow-band [SII] image of the core of the HH 34 flow showing HH 34S (previously known as HH 34) in the bottom center of the figure and the fainter bow shock HH 34N towards the top. The HH 34 jet (see also the HST image in Fig. 4) emanates from an embedded infrared source, HH 34IRS, at its northern apex and points towards HH 34S. Both bow shocks are approximately $100''$ away from the infrared source. The HH 34 flow is itself much more extensive (Bally and Devine 1994) as already hinted at by this frame since additional parts of the flow can be seen beyond HH 34N and S. This image, courtesy of J. Eislöffel and R. Mundt, was taken with the 3.5-m telescope on Calar Alto, Spain. North is to the top and east is to the left.

Heathcote and Reipurth (1992) used radial velocities measurements, in combination with proper motion data to infer the actual velocities of HH 34 N and S. They found that both bow shocks were receding from the source at approximately 330 km s^{-1} along an axis which makes an angle of approximately 20–30° to the plane of the sky. If HH 34 S is a bow shock moving into a stationary ambient medium, then, given its high velocity, one would expect correspondingly high shock velocities at its apex. In fact the shock velocities inferred from various emission line ratios appear to be relatively low (approximately 160 kms^{-1}). Thus we are forced to conclude that HH 34 S, and presumably HH 34 N, are in fact advancing into an medium that is already moving forward. This is consistent with the finding that the HH 34 flow is much larger than previously thought (Bally and Devine 1994). It also means that both HH 34 N and S are essentially large internal working surfaces and do not mark the ends of the flow.

3. THE MORPHOLOGY OF YSO JETS

The HH 34 jet, our archetype, also illustrates a recurrent theme amongst YSO jets in that they usually consist of a series of sometimes quasi-periodically spaced knots. Although there is some discussion about their origin, recent HST images leave no room for doubt that most knots are internal working surfaces (see §6) as originally proposed by Raga and Kofman (1992). That said, alternative explanations have been put forward such as the non-linear growth of Kelvin-Helmholtz (K-H) instabilities (see, for example, Micono Chapter 21 and Massaglia Chapter 20) and a K-H model may be appropriate is some cases.

A number of HH and molecular (see below) jets are seen to wiggle like HH 30 (e.g., López et al. 1995; Mundt et al. 1990) and RNO 15 (Davis et al. 1997a) but the origin of these wiggles is a good deal less certain than that of the knots. The wiggles may be due to precession or perhaps [magneto-]hydrodynamic instabilities could play a rôle. Certainly the HH 34 super-flow (Bally and Devine 1994) exhibits S-shaped symmetry indicative of a slow drift in the direction of the ejection axis by 5-10° over a period of 10^4 years or so. Similar drifts are seen in the RNO 43 flow (Padman, Bence and Richer 1997). In these cases we are almost certainly dealing with some form of precession possibly caused by a gradual change in the direction of the angular momentum of the accreted material. If the wiggles on smaller spatial scales are due to, for example, the helical K-H instability then this strikes me as something which should in principle be testable by measuring how the wiggles evolve with time given that we know the jet parameters.

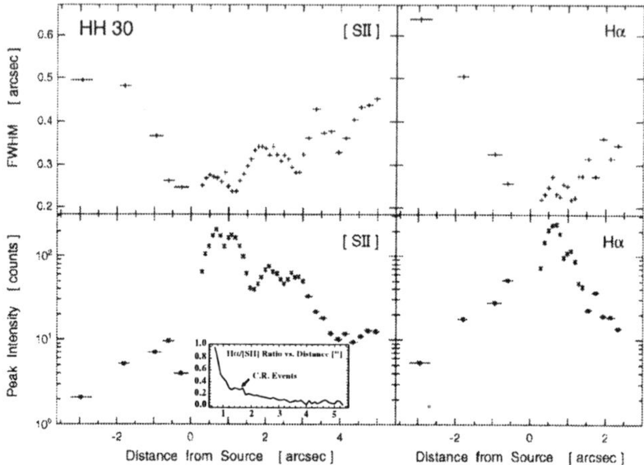

FIGURE 2. Plots (top) of how the diameter (Full Width Half Maximum) of the HH 30 bipolar jet (see also Fig. 5) varies with distance from the source based on HST imaging in the red [SII] doublet and Hα line (from Ray *et al* 1996). The bottom graphs show the corresponding variations in relative intensity i.e. they show the first few knots. Note how the diameter is still broad ($\approx 0''.25$) at a distance of $0''.3$ from the source (see text) and has not changed appreciably from its value at $3''.0$.

A rather curious characteristic of YSO jets (unique as far as I know since it does not seem to have been observed amongst their extragalactic counterparts) has emerged from an examination of how their diameters vary with distance from the source. Mundt, Ray and Raga (1991) noticed from ground based studies (and more recently this result has been confirmed by the HST (Ray et al. 1996) that the diameter of YSO jets decrease only gradually as one gets closer to the source in such a way that the effective opening angle, defined by the diameter (Full Width Half Maximum) of the jet divided by the distance to the source, increases. For example, in the case of HH 30 (Figs. 2 and 5), the jet diameter changes very little between $3''.0$ and $0''.3$ from the source remaining close to $0''.25$. There are problems, even in an ideal case like HH 30 with optically imaging the jet even closer still (in particular there are difficulties removing the effects of contaminating light from the star) but such observations are at least consistent with current ideas as to how magnetic fields play a fundamental role in collimating YSO jets (see, for example,

Camenzind, Chapter 1). In particular, most MHD models predict an initially poorly collimated flow within a few tens of AU from the source, which is then subsequently collimated into a more 'column-like' structure as magnetic hoop stresses become important.

4. SOME NUMBERS

As stated in the Introduction, one of the beautiful features of YSO jets is that many of their physically important parameters can be derived through spectroscopic studies. The observed strengths of the various emission lines tell us not only that the knots must be radiative shocks but we also know, for example, that the typical temperatures of the knots are around 5×10^3–10^4 K. Combination of proper motion data and radial velocity measurements imply jet velocities usually in the range 200-1000 km s^{-1}. There is a tendency for the jet velocity to scale with the luminosity (or more probably the gravitational escape velocity) of the source as one might expect (e.g., Mundt and Ray 1996). With these figures one can determine typical hydrodynamic Mach numbers for YSO jets and the values found are very high ($M_{jet} \approx 20$–100). Even in the absence of any confining medium, and ignoring any rôle magnetic fields might play, these jets should remain highly collimated, as is observed, once focused.

Electron densities can be derived using, for example, the ratio of the [SII]$\lambda\lambda6716, 6731$ lines. Characteristic N_e values around 10^2–10^4 cm^{-3} are found. The typical neutral fraction in jets appear to be quite high and may be more than 90% (e.g., Bacciotti, Chiuderi and Oliva 1995). Jet mass fluxes depend, as one might expect, on the luminosity of the YSO, and are typically in the range 10^{-8}–10^{-6}M$_\odot$yr^{-1} (Edwards, Ray and Mundt 1993).

Turning now to the sizes of HH flows, we have already stated, with reference to HH 34, that this flow is much larger than previously thought and extends for a least a few parsecs. Originally (e.g., Mundt, Brugel and Bührke 1987) it was thought that most HH flows were much smaller; now, however, with the growing sizes of CCDs, average estimates of the total flow lengths has also increased [!] (Bally and Devine 1994; Ogura 1995; Eislöffel and Mundt 1997) and projected total sizes of several parsecs are becoming the norm. Even a decade ago (Ray 1987) it was realized that some flows, like RNO 43 (HH 245, HH 243) were more than a parsec in length and others, like the HH flow from Z CMa (Poetzel, Mundt and Ray 1989) stretched even further. In retrospect this is not too surprising. Statistical studies (e.g., Mundt et al. 1987 and Parker, Padman and Scott 1991) indicated that the outflow phase for solar-like stars typically persists for at least a few times 10^4 to 10^5 years. As Padman et al. (1997) point out, a 200 km s^{-1} jet will travel 20 pc in 10^5 years and, in fact, one would expect, given the sizes of molecular clouds, that in many cases these HH flows have long escaped from their parent cloud making it difficult to see their true terminal working surfaces.

5. HH JETS AND MOLECULAR OUTFLOWS

It would be impossible in this short paper to give an in-depth study of molecular outflows and the reader is instead referred to the excellent review paper by Bachiller (1996). Here I wish to concentrate on recently developed 'unified' models according to which the much slower CO molecular outflows are driven, somehow, by an underlying HH-type jet. In a number of cases where CO outflows are present (e.g., HH 46/47 Eislöffel et al. 1994) the HH jet is seen either optically or in the near-infrared, but more often than not, at least for the most embedded flows, we can only infer its presence. There are several reasons why it is thought CO outflows are actually driven by HH jets. Firstly, HH jets seem to have sufficient momentum to do so (it is generally agreed that these flows are not energy driven). Originally the neutral fractions of HH jets were grossly underestimated (e.g., Mundt, Brugel and Bührke 1987). This, in turn, lead to underestimates of the momentum fluxes of these jets and dismissal of the idea that they could drive any associated molecular outflow. We now know this to be wrong: the upwardly revised values for $\dot{M}_{jet}$ have dispelled this problem. Another observation that points to HH jets being the power-house for the CO outflows is that many molecular outflows, especially when observed with large radial velocity offsets, appear highly collimated (e.g., NGC 2024, see Richer, Hills and Padman 1992). Such flows (Bachiller 1996) exhibit high velocity components ($\approx$ 30–50 km s^{-1}) towards the flow axis but at lower velocities look more and more like the 'standard' CO flows (e.g., L1551-IRS5, see Moriarty-Schieven and Snell 1988). It is not clear whether these very high velocity 'CO jets' are largely entrained material that has been accelerated to such velocities by an underlying HH jet or, and this seems less likely, whether they represent the more neutral component of the latter (Bachiller 1996). Certainly the observation (Padman et al. 1997) that the collimation of these flows increases monotonically with radial velocity offset, while the apparent mass *decreases*, suggests that they are the result of entrainment. Moreover the velocities of the 'CO jets', although high are not as high as those of the HH jets, so again this points to the 'CO jets' representing initially accelerated ambient gas. Moreover there is another reason to think that none of the CO is directly accelerated at the source. This is perhaps most clearly illustrated when we look at bipolar HH flows such as HH 46/47 which appear monopolar in CO. Here the source is located at the edge of a dark cloud and, interestingly, the CO molecular outflow is seen only on the side where the HH jet flows *into* the parent cloud. This tells us that the CO outflow is entrained ambient material from the surrounding cloud and does not originate at the star itself.

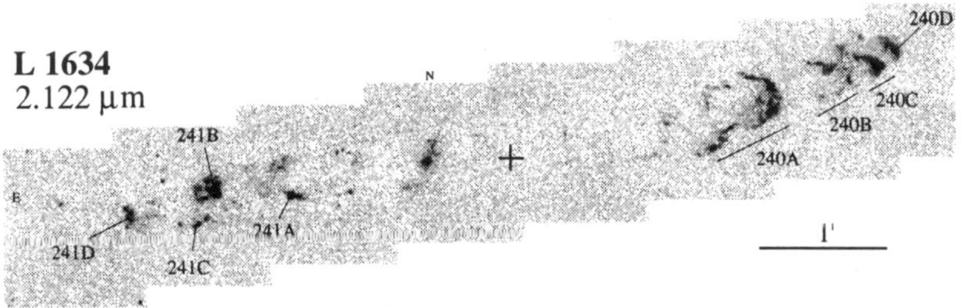

FIGURE 3. The embedded Lynds 1634 outflow as seen in the 2.12μm line of H_2 from Davis et al. (1997b). The far-infrared source is located towards the center of the frame and is marked with a cross. Note the extensive bow shocks that are seen in both outflow directions. North is up and east is to the left.

So far, however, we have not said anything about how such a 'unified' model for HH jets and molecular flows might work. In this regard the various emission lines of H_2 have proven to be an invaluable tool. Several studies, see Eislöffel (1997) for a general review, have shown that outflows, particularly highly embedded ones, can be traced in shocked H_2 emission (see, for example, Fig. 3) although we do not, at this stage know whether the emission is due to discontinuous jump (J) -type or continuous (C) -type magnetic shocks. Often bright H_2 bows are seen (Fig. 3). In any event, two principal entrainment mechanisms have been proposed to accelerate the ambient material i.e. to transfer momentum from the HH jet to its surroundings (see, for example, Bachiller 1996). The first is 'prompt entrainment', which occurs at the head of the jet, or at least at the internal working surfaces, and the second is 'steady state entrainment' which occurs along its sides and involves turbulent mixing with the external material, most likely through Kelvin-Helmholtz instabilities. Over a decade ago, De Young (1986) showed that, at least in the case of adiabatic jets, the first process dominates for high Mach num-

bers ($M_{jet} \geq 5$), whereas 'steady state entrainment' is only important
if jets are mildly supersonic. As HH jets appear to have very high
Mach numbers, it seems much more likely that the 'prompt entrainment' mechanism (e.g., Raga and Cabrit, 1993; Chernin et al. 1994)
is important in actual outflows and this is supported by the H_2 observations. If one overlaps CO maps with H_2 images, it is immediately obvious (e.g., Davis and Eislöffel 1995) that there is an intimate
connection between the two and that in almost all outflows the CO
peaks upstream of the H_2 bows. This clearly suggests that the CO
is promptly entrained gas. Moreover recent observations by Davis et
al. (1997a) of the RNO15-FIR flow show that the observed H_2 bows
coincide with peaks in the momentum, and mass per unit length, of
the CO flow again implying prompt entrainment.

6. HUBBLE SPACE TELESCOPE OBSERVATIONS OF YSO JETS

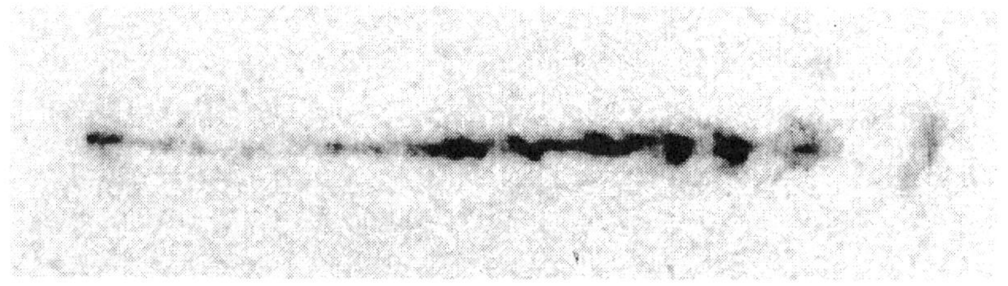

FIGURE 4. HST [SII]$\lambda\lambda$6716,6731 image of the HH 34 jet in which several
individual knots are resolved into bow shocks. Faint shocks are also seen
to propagate into the medium surrounding the jet. From such images it
would appear that temporal variability in the outflow, which in turn leads
to the formation of 'internal working surfaces' is responsible for most of the
observed knots. The jet has been oriented so that the source is on the left
and the flow is in the horizontal direction.

As stated earlier, the HST has proved to be invaluable in the
study of YSO jets. In particular, it has helped to a large degree to
resolve the controversy surrounding the origin of jet knots and has
allowed us to probe the innermost tens of AU from the source in an
effort to understand the actual origin of the jets themselves. Partly
the reason why the superior resolution of the HST has prove to be
important is because most jets have typical angular diameters around
$0''.5$–$1''.0$ and therefore a lot of interesting sub-structure happens to
be on this or smaller scales. To illustrate the point, Fig. 4 shows a
HST image of the HH 34 jet taken through a red [SII]$\lambda\lambda$6716,6731

filter (based on archive data obtained by the Wide Field Planetary Camera 2 Team, see also Ray et al. 1996).

In Fig. 4 it can immediately be seen that several of the knots are resolved into 'mini-bow shocks' supporting the idea that these knots arise as a result of temporal variations in the outflow and that they are caused by faster jet material catching up with previously ejected slower gas. Moreover we see that the bows all point away from the source implying that there is a shear layer across the jet with the jet velocity peaking towards the central axis.

Another important 'first' for the HST is that it has allowed us to resolve the actual cooling zones of the internal jet shocks, i.e. the knots. In the case of a bow shock, the shock velocity is obviously highest at the apex and so we would expect the highest ionization should occur there rather than in the wings. Back from the apex we would expect a mixture of higher excitation lines like $H\alpha$ as well as lower excitation ones such as [SII]$\lambda\lambda6716,6731$. In, for example, both the HH 34 jet and the HH 111 jet (Reipurth et al. 1997), the $H\alpha$ emission is seen largely to be confined to the jet axis while the [SII] emission comes from the wings and the region trailing behind the apex. This resolution of the cooling zone is as predicted by shock theory. In a number of cases the Mach disc is also observed within the bow shock (e.g., Reipurth et al. 1997) and typically has a diameter of around a quarter of the bow shock width, as expected on the basis of numerical simulations, for example, Stone and Norman 1993).

Although internal working surfaces may be an explanation for many of the observed knots in YSO jets, this may not be the case for all. Take for example the HH 30 flow (Fig. 5). Here none of the well resolved knots close to the source appear to be bow-shaped (Ray et al. 1996) although admittedly one would not expect this for internal working surfaces if the jet has a 'top-hat' type velocity profile. More importantly a monotonic decrease in the [SII]/$H\alpha$ ratio is observed for the first few knots, whereas this ratio might be expected to vary at the knots if material is being re-shocked. The knots in this case could be mild condensations in the flow.

The HST images of HH 30 (Fig. 5, Burrows et al. 1996 and Ray et al. 1996) are interesting for another reason. From ground-based studies (Mundt et al. 1990) it was already known that this flow is virtually in the plane of the sky. As can be seen from Fig. 5, the HH 30 bipolar jet emerges perpendicular to two cusp-shaped reflection nebula. These cusp-shaped nebula are illuminated by scattered light from the source which is deeply embedded (Burrows et al. 1996).

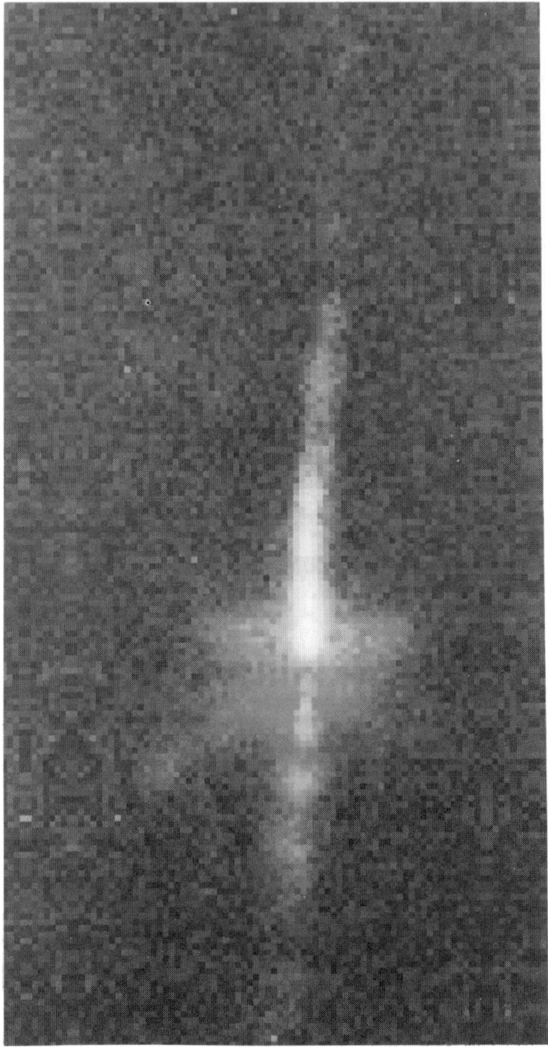

FIGURE 5. HST image of HH 30 based on data from Ray et al. (1996). The jet has been oriented so that the blue-shifted flow points almost upwards. Blue light represents continuum emission, which in this case is scattered starlight from the 'top' and 'bottom' of the disk. The source itself is highly embedded and the disk can be seen in silhouette as the dark lane bisecting the nebular cusps. Red light is emission from the red [SII] doublet and green represents Hα. Note the asymmetry in excitation conditions in the flow and the counterflow. Image reconstruction done by C.R. O'Dell and S.V.W. Beckwith (see Color Plate 3).

A dark lane can be seen between the two cusps, which is presumably the disc around this star seen edge-on, as expected for a flow that is almost in the plane of the sky. In other cases, e.g. HL Tau (Ray et al. 1996; Stapelfeldt et al. 1996) the geometry is more complicated and, while the presence of a disc can be inferred, it does not manifest itself so clearly.

7. CLOSE TO THE SOURCE

So far this review has concentrated almost exclusively on the extended structure of YSO jets, i.e. on angular scales of several arcseconds which, for nearby star forming regions like Taurus-Auriga, correspond to distances of several thousand AU. We have said very little about what is observationally found close to the source. Ultimately, however, it is this very region that we must study if we are to obtain clues as to how YSO jets form in the first place. Of course many jet sources are highly embedded and so it is often impossible to probe their immediate environments, at least at optical wavelengths. That said there are a number of optically visible classical T Tauri stars and Herbig Ae/Be stars which are jet sources. Often these jets appear to extend for only a few arcseconds (hence their nickname 'microjets', see Solf 1997) but such flows can, I believe, help us understand the YSO jet formation process.

At the same time it should be emphasized that even in those cases where the jet can be traced back optically to the source, there are a number of problems in studying it. For example the contaminating effects of light from the YSO has to deal with. Moreover scattered light from reflection nebula can cause difficulties even when narrow-band imaging is used. To a large degree these problems can be overcome by the use of long-slit spectroscopy (Solf and Böhm 1993; Böhm and Solf 1994). With intermediate dispersions spectrograms, the continuum light from the star can be subtracted very effectively; allowance can even be made for stellar absorption lines using a suitable template spectrum. The resultant spectra for individual lines are then effectively 1-D position velocity diagrams and the various line ratios can give us information on physical conditions in the emitting gas.

When one looks at a typical T Tauri or Herbig Ae/Be star jet source, its spectrum normally shows at least two emission line velocity components and sometimes more. The flow close to the source is perhaps best seen in forbidden transitions, e.g. [OI]λ6300 or [SII]$\lambda\lambda$6716,6731, as the Balmer lines may have some contribution from the star itself. The high velocity component, or HVC as it is often known, usually has a velocity of more than a hundred km/s and is can be identified with the extended jet where seen (Kwan and Tademaru 1995). The low velocity component, or LVC, on the other hand has a velocity closer to the systemic velocity of the star. Both components, however, are normally blue-shifted, at least close to the star. This asymmetry is thought to be due to obscuration caused by a circumstellar disc which hides the red-shifted counterflow (Edwards, Ray and Mundt 1993; Corcoran and Ray 1997).

It has been suggested that the LVC is a disc wind (Kwan and Tademaru 1995) although its origin is not well understood. In any event the gas in the HVC and LVC usually have quite different excitation conditions and also different spatial properties (see, for example, Hirth, Mundt and Solf 1997). Normally the LVC is of much higher density, at least for classical T Tauri stars, and of lower excitation than the HVC. Both the strength of the LVC and the HVC appears to scale with various measures of accretion (e.g., Calvet 1997) and this clearly points to a common origin for both components. Any model which seeks to explain the origin of the jet (HVC) emission must simultaneously, I believe, be capable of explaining the origin of the LVC as well. The latter is something which is often ignored by theorists although some promising dual component models have been proposed (e.g. Goodson, Winglee and Böhm 1997). It has even been argued that there is no real distinction between the two components and that they are merely different facets of the same flow (Calvet 1997). Certainly understanding the relationship between the LVC and HVC is a problem that deserves more attention in the future.

Finally there is general, although by no means universal (e.g. Frank and Mellema 1996), agreement that magnetic fields play an important rôle in the collimation of YSO jets (e.g. Camenzind, Chapter 1). In this regard it is interesting that until recently there has been no unambiguous evidence for magnetic fields in YSO outflows. Using the Multi-Element Radio Linked Interferometer Network (MERLIN), however, Ray et al. (1997) found that the radio emission from T Tau S, the embedded infrared companion to T Tau, is circularly polarized and extended on scales of 20–40 AU. The extension was observed roughly in the supposed direction of the outflow from this star and the polarization was found to be of opposite helicity in the flow and the counterflow. In this case the radio emission is almost certainly gyrosynchrotron radiation implying the presence of mildly relativistic electrons which are probably shock accelerated. Given the large degree of circular polarization present, and using reasonable estimates of the parameters involved, one can infer field strengths of at least a few Gauss at 20-40 AU. Such field values are quite large suggesting, as most theorists presume, that magnetic fields are dynamically important in YSO jets.

Acknowledgements

The author wishes to acknowledge the very kind support and hospitality shown to him by the Dipartimento di Fisica Generale dell'Università di Torino and the Osservatorio Astronomico di Torino during his visit to Turin.

REFERENCES

Bacciotti, F., Chiuderi, C., & Oliva, E., 1995, *A&A*, **296**, 185.

Bachiller, R., 1996, *Ann. Rev. Astron. Astrophys.*, **34**, 111.

Bally, J., & Devine, D., 1994, *Ap.J. Lett.*, **428**, L65.

Böhm, K.H., & Solf, J., 1994, *Ap.J.*, **430**, 277.

Bührke T., Mundt R., & Ray T. P., 1988, *A&A*, **200**, 99.

Burrows, C.J., *et al*, 1996, *Ap.J.*, **473**, 437.

Calvet, N., 1997, in *Herbig-Haro Flows and the Birth of Low Mass Stars*, eds. B. Reipurth and C. Bertout, Kluwer Academic Press, p.417.

Chernin, L., Masson, C., Gouveia Dal Pino, E.M., & Benz, W., 1994, *Ap.J.*, **426**, 204.

Corcoran, M., & Ray, T.P., 1997, *A&A*, **321**, 189.

Davis, C.J., Eislöffel, J., Ray, T.P., & Jenness, T., 1997a, *A&A*, in press.

Davis, C.J., Ray, T.P., Eislöffel, J., & Corcoran, D., 1997b, *A&A*, in press.

Dopita, M.A., Schwartz, R.D., & Evans, I., 1982, *Ap.J. Lett.*, **263**, L73.

Edwards, S., Ray, T.P., & Mundt, R., 1993, in *Protostars and Planets III*, eds. E. Levy & J. Lunine, University of Arizona Press, p.567.

Eislöffel, J., 1997, in *Herbig-Haro Flows and the Birth of Low Mass Stars*, eds. B. Reipurth & C. Bertout, Kluwer Academic Press, p.93.

Eislöffel, J., Davis, C.J., Ray, T.P., & Mundt, R., 1994, *Ap.J. Lett.*, **422**, L91.

Eislöffel, J., & Mundt, R., 1997, *A.J.*, in press.

Frank, A., & Mellema, G., 1996, *Ap.J.*, **472**, 684.

Goodson, A.P., Winglee, R.M., & Böhm, K.H., 1997, *Ap.J.*, in press.

Haro, G. 1952, *Ap.J.*, **115**, 572

Hartigan, P., Raymond, J., & Hartmann, L., 1987, *Ap.J.*, **316**, 348.

Herbig, G.H., 1951, *Ap.J.*, **113**, 697.

Herbig, G.H., 1974, Lick Observatory Bulletin No. 658.

Hirth, G., Mundt, R., & Solf, J., 1997, *A&A Suppl.*, in press.

Kuhi, L.V., 1964, *Ap.J.*, **140**, 1409.

Kwan, J., & Tademaru, E., 1995, *Ap.J.*, **454**, 382.

Lopéz, R., Raga, A., Riera, A., Anglada, G., & Estalella, R., 1995 *MNRAS*, **274**, L19.

Moriarty-Schieven, G. H., & Snell, R.L., 1988, *Ap.J.*, **332**, 364.

Mundt, R., 1986, *Canadian J. Phys.*, **64**, 407.

Mundt, R., & Fried, J.W., 1983, *Ap.J. Lett.*, **274**, L83.

Mundt, R., Brugel, E.W., & Bührke, T., 1987, *Ap.J.*, **319**, 275.

Mundt, R., Bührke, T., Fried, J. W., Neckel, T., Sarcander, M., & Stocke, J., 1984, *A&A*, **140**, 17.

Mundt, R., & Ray, T.P., 1994, in *The Nature and Evolutionary Status of Herbig Ae/Be Stars*, eds. P.S. Thé, M.R. Perez & E.P.J. van den Heuvel, P.A.S.P. Conf. Ser., p.237.

Mundt, R., Ray, T.P., Bührke, T., Raga, A., & Solf, J., 1990, *A&A*, **232**, 37.

Mundt, R., Ray, T.P., & Raga, A.C., 1991, *A&A*, **252**, 740.

Norman, C., & Silk, J., 1979, *Ap.J.*, **228**, 197.

Ogura, K., 1995, *Ap.J. Lett.*, **450**, L230.

Osterbrock, D.E., 1958, *PASP*, **70**, 399.

Padman, R., Bence, S., & Richer, J., 1997, in *Herbig-Haro Flows and the Birth of Low Mass Stars*, eds. B. Reipurth & C. Bertout, Kluwer Academic Press, p.123.

Parker, N.D., Padman, R., & Scott, P.F., 1991, *MNRAS*, **252**, 442.

Poetzel, R., Mundt, R., & Ray, T.P., 1989, *A&A*, **224**, L13.

Raga, A.C., & Cabrit, S., 1993, *A&A*, **278**, 267.

Raga, A.C., & Kofman, L., 1992, *Ap.J.*, **386**, 222.

Ray, T.P., 1987, *A&A*, **171**, 145.

Ray, T.P., Mundt, R., Dyson, J., Falle, S.A.E.G., & Raga, A., 1996, *Ap.J. Lett.*, **468**, L103.

Ray, T.P., Muxlow, T.W.B., Axon, D.J., Brown, A., Corcoran, D., Dyson, J., & Mundt, R., 1997, *Nature*, **385**, 415.

Reipurth, B., 1994, A General Catalogue of Herbig-Haro Objects, electronically published using anonymous ftp at ftp.hq.eso.org, directory /pub/Catalogs/Herbig-Haro.

Reipurth, B., Bally, J., Graham, J.A., Lane, A.P., & Zealey, W.J., 1986, *A&A*, **164**, 51.

Reipurth, B., Hartigan, P., Heathcote, S., Morse, J.A., & Bally, J., *A.J.*, in press.

Richer, J.S., Hills, R.E., & Padman, R., 1992, *MNRAS*, **254**, 525.

Schwartz, R.D., 1975, *Ap.J.*, **195**, 631.

Schwartz, R.D., 1977, *Ap.J. Lett.*, **212**, L25.

Schwartz, R.D., 1978, *Ap.J.*, **223**, 884.

Solf, J., 1997, in *Herbig-Haro Flows and the Birth of Low Mass Stars*, eds. B. Reipurth and C. Bertout, Kluwer Academic Press, p.63.
Solf, J.,& Böhm, K.H., 1993, *Ap.J. Lett.*, **410**, L31.
Stone, J.M., & Norman, M.L., 1993, *Ap.J.*, **413**, 210.
Strom, S.E., Strom, K.M., Grasdalen, G.L., Sellgren, K., & Wolff, S., 1985, *A.J.*, **90**, 2281.

CHAPTER 17

PROTOSTELLAR JETS AND MOLECULAR OUTFLOWS: THE CHICKEN OR THE EGG?

R. N. HENRIKSEN

Astronomy Group, Department of Physics, Queen's University at Kingston, Ontario, K7L 3N6, Canada

One knows now that, at least in some of the celebrated objects containing Herbig-Haro components, high-velocity, well-collimated, jets are associated with the lower velocity molecular outflow sources. But is it a hidden high velocity stellar 'wind' that is collimated to these jets that, either directly or subsequently (by shocks and entrainment associated with the jet), establishes the outflow in the ambient medium; or does the outflow have a separate existence that itself favors the creation of jets? Does protostellar evolution result in these different active elements being formed at different times? Is the paradigm of a steady stellar wind really relevant? In this paper I address these questions based on relevant published observations and theories, and on simple physical considerations. It is argued that we may not have considered sufficiently the role of violent magnetic flares in our models at present.

1. INTRODUCTION

In this chapter I will argue that the questions posed in the abstract may not have the conventional answers. Rather these may require distinguishing the proto-stellar evolutionary epochs at which the molecular outflows and the high velocity optical jets *first* occur, while regarding the low speed and high speed molecular outflow to be simultaneous with the growth phase of the star. Because of the large difference in dynamical times between the molecular outflows and the optical jets in the same volume of space, these can appear to be spatially coincident even when the launch times differ by a factor as large as the ratio of the respective velocities. I shall argue below for this long term variability and emphasize the potential importance of a shorter time-scale 'flaring' type variability .

2. OF EGGS AND CHICKENS AND MUTATION

Molecular outflows and collimated high velocity jets are clearly part of the sequence of protostellar evolution. It is useful to begin any discussion of these phenomena by recalling the available energy sources. Gravitational energy must dominate the immediate post-stellar-core formation 'star-building' or accretion phase almost by definition. It may in part be thermalized, magnetized or 'kinetized' but it remains the ultimate source. Eventually deuterium core and then deuterium shell burning (for stars more massive than about 2 $M_\odot$ (Palla and Stahler 1990) provide independent sources of energy and then finally, with the termination of the accretion phase, the star ignites hydrogen in the core and settles onto the main sequence. However the gross accretion luminosity (ignoring the energy converted for the moment) dominates the fusion power in the D burning phases for the anticipated accretion rates of $\approx 10^{-5}$ $M_\odot$ yr^{-1}.

All models of this early 'star-building' phase thus appeal to the accretion power as the prime mover (see e.g. Henriksen, André and Bontemps, 1997 for a discussion of the power available). However the evidence of the class 0 protostellar sources, which are exactly in this star-building phase, (e.g. Bontemps et al., 1996) is that this phase (in at least some star forming regions) is an even *more* prodigious generator of molecular outflow than are the older class I sources. From this we conclude that infall and molecular outflow must occur during the class 0 stage, and that there is a singularly efficient mechanism for creating the outflows from these sources. This latter deduction already suggests that there may be a similar optimal phase for the optical jets. The former deduction raises the question of the efficiency of the outflow, since clearly not all the mass that falls onto the stellar core can be re-ejected at high velocity.

This last question of outflow efficiency, or of how the infalling material is distributed between that stored in orbit, that incorporated into the star, and that re-ejected, is intimately linked to the topology and morphology of the global circulation of material. Very general arguments following from the joint action of gravity and inertia yield 'discs' surrounding the stellar core, but such discs may resemble more spheres with axial cones removed than the traditional thin disc of cold gas and dust. For it is possible that these 'discs' are the result of a delicate dynamical balance between magnetic fields, gravity, inertia and thermal pressure as argued recently by Henriksen and Valls-Gabaud (1994, HVG) and especially Fiege and Henriksen (1996a,b, FHa,b).

With only this much common ground, we can already discuss qualitatively the question embodied in the title of this paper thus:

i) If the optical jets are driving the outflow in the class 0 phase (but are nevertheless largely unseen), then they should arise close to the central protostar where the accretion power is maximal and *the free-fall or Keplerian speed is typical of the maximum observed jet velocity*. For all jet launch models involve the passage of the flow through a critical point where the gravitational energy must roughly balance the driving energy, be it thermal or magnetic. The nature of these critical points in different outflow self-similar geometries has recently been discussed powerfully by Tsinganos et al. (1996). Moreover to constitute a real explanation of the *class 0 phenomenon* an optical jet must also be created as part of the accretion flow that forms the star.

ii) Recently numerical simulations by Ouyed, Pudritz and Stone (1997) have treated the magnetic launch (rather than an assumed steady state) of an inner disc wind that becomes a self-collimated jet. This is a mechanism that satisfies the constraints of paragraph 1 except that it is really not incorporated into the global infall/outflow. That is, the disc is not active and no material flows inward through it. Thus we still do not know that this is an inevitable consequence of the stellar growth phase, although this work does demonstrate clearly that the Blandford-Payne (1982) mechanism can work. Steady state self-consistent models of this type do exist (Ferreira and Pelletier, 1995) as well as an interesting class of disc models where the launch is *at* the disc Alfvénic point (e.g. Najita, 1995) but as yet these have not been united with the surrounding molecular outflow.

iii) The subsequent driving of the less well collimated molecular outflows presents a crucial problem for the jet models as prime mover. As discussed lucidly by Bachiller (1996) the transfer of momentum must be accomplished either at the bow shock(s) of the jet or by turbulent entrainment along the sides (simple analytic models were given by Henriksen, 1987). The models (e.g. Raga and Cabrit, 1993) have difficulty establishing both the transverse dimensions of the outflow and the correct velocity field. (Somewhat paradoxically, the difficulty in maintaining the non-linear stability of large-scale jets revealed at this meeting by Massaglia (Chapter 20) and Bodo (Chapter 15) may assist in diffusing the outflow over the ambient medium.) Moreover it is not certain that the observed jets transport enough momentum to provide the flux in CO, and in any case it requires the fastest observed optical jets at low fractional ionization, which are not always known to be present. However they do seem to be more plausible than the wind driven models (Masson and Chernin, 1992) unless these are given an already small bipolar momentum X-section, which rather begs the question.

We are left with the conclusion that despite the enormous effort expended, we still do not have a model of global infall/outflow in the class 0 star-building phase that manifestly:

a) focuses most of the outgoing momentum into a well collimated jet near the stellar core *in a dynamical fashion consistent with the global flow*,

b) produces the observed mass and momentum outflow in real space and in phase space.

Thus if we identify such a jet as the 'egg' (with the molecular outflow as the 'chicken'), it is not yet evident that it comes first!

A different approach has been followed in the studies of HVG and FHa,b. There the molecular outflow is regarded as a natural consequence of the circulation established by the collapse of the pre-stellar cloud. 'Cored apple' type distributions of circum-protostellar gas, a convective pattern of infall/outflow, self-consistent axial collimating magnetic fields and rotation, and an outflow velocity that increases toward the axis of the distribution as required by the observations; are all produced as a consequence of the model rather than by ad hoc considerations. The magnetic field/stream lines are quadrupolar in the poloidal plane, but the flow rises over the poles in a 'tornado-like' self-collimated structure. The highest velocity outflow is found along the axes of these 'tornados' basically due to the low density there. The flow is essentially 'rebounding' off the toroidal pressure distribution identified with the cored apple 'disc' to 'miss' the stellar core in a non-Keplerian orbit at essentially the free-fall speed. A moderate amount of heating supplied at this turning point allows the outflow to possess finite velocities at infinity. The models published have *no* magnetic driving (zero Poynting flux) in order to test for its necessity, but it can readily be incorporated (Henriksen, in preparation).

In FHb the line shapes, velocity cut intensity maps and integrated intensity maps have been calculated based on the model of FHa. The fit to the observations is very good, with in particular the conical structures seen dramatically in such sources as L1448 (e.g. Bachiller, 1996) reproduced. Since all of the ambient gas is set into motion by the gravitational collapse, none of the problems of momentum transfer discussed above present themselves. The 'chicken' then functions pretty well as a chicken and it lays a moderately high velocity 'egg'. That is jet velocities up to 50 km s^{-1} are easily observable in the model, but beyond that it is necessary to approach the stellar core ever more closely at the turning point (FHa), and it is not clear that these highly collimated streams would be observed.

All is not yet ideal in this type of model. The origin of the required quadrupolar steady state is left unknown by the model. The accretion rate onto the star is purely due to its geometric X-section as it intercepts the quadrupolar stream lines, and in any case only a point non-accreting star is strictly consistent with the current model. This is a requirement of the steady state assumption and I feel that this must ultimately be abandoned as discussed below. The very high speed optical jets may be beyond the reach of the model as it now exists.

Shocks exist naturally in this model, not only as the bow shock of the high speed axial flow, but more likely as bipolar conical structures containing the axis. Since the type 1 solutions of FHa are always super Alfvénic, these shocks will be approximately described by the ordinary equations of hydrodynamics. The type 2 solutions seem to contain a fast magnetosonic critical point and may therefore possess more complicated MHD shocks. Such shock structures provide a natural explanation for the excited H_2 sheath emission in such sources as HH 211 (class 0; Bachiller, 1996).

In addition, some recent numerical simulations by Toropin, Saveljev and Chechetkin (1996,TSC) remove at least some of the mystery of a possible origin of the FH steady state. These authors attempt to find a global solution to the infall/outflow numerically by evolving the system from an initial state. This consists of a rotating disc of radius $\approx$ 40 AU, penetrated normally by an ordered magnetic field. Onto this disc is allowed to fall a kind of Bondi spherical accretion flow. In time (about 100 years of real time) the system develops approximately quadrupolar stream lines, conical shock structure and axial outflow with Mach numbers up to about 6 (about 40 km s^{-1}). They find that about 20% of the infalling material is re-ejected. Generally the development looks and acts remarkably like the FH steady state with the addition of the anticipated shocks. The simulations show in addition that the Ampèrian force is small compared to pressure gradients. Thus in fact one might expect these results to resemble the results of similar calculations made *without* the magnetic field by Clarke et al. (1985) and more recently by Chen et al. (1997). Indeed the general features, the quadrupolar flow, the shock structure and the thick disc are all confirmed by these calculations.

One of the more interesting facts that follows from these various simulations is the development of intense vortex rings near the inner boundary of the disc, and indeed a complicated overall structure of vorticity and magnetic intensity in this region (TSC). We shall recall this in our speculations below.

All of this is making a fair case perhaps for at least the *molecular* 'chicken' and the 'egg' having evolved together in good Darwinian fashion. However none of the numerical simulations, and likely not the FH type models as we have seen, yet contain anything that resembles the extreme HH type jets. I begin to wonder whether these might be a later phenomenon in the protostellar evolutionary sequence. That is, perhaps the chicken and egg paradigm for low-speed infall/outflow and high speed collimated axial outflow is overtaken by a now massive young star in the Herbig Ae/Be phase asserting itself through an active corona and a vigorous stellar wind. The molecular circulation still remaining after the star building phase would simply serve as a perfectly formed collimator for the optical jets. The short dynamical time of the optical jets permits them to overtake the fossil outflows and so to lead them spatially even though they may trail in time. The system then is subject to mutation on the star formation timescale, and molecular outflows might be regarded as the 'missing link' between protostars and pre-main-sequence stars and their jet activity.

In addition to this evolutionary variability, I believe that we have strong evidence for shorter time scale activity in the clumpy nature of some jets and from the concentrations of vorticity and magnetic intensity seen in the simulations. I shall suggest in the next section a novel way of thinking analytically about these possibilities.

3. SOURCES OF VARIABILITY

In this section I summarize some speculations on the nature and effects of variability.

The argument for jet/outflow variability on the protostellar time scale is basically the changing relative importance of accretion power and the luminosity of the central source. We know that convective stars (those of mass less than ≈ 2 $M_\odot$, Palla and Stahler 1990) produce violently active coronae, and in fact even the sun is weakly bipolar source. For the dramatic Coronal Mass Ejections (CME) occur with velocities from 500 to 1000 km s^{-1}, with the higher velocities occurring over the poles as revealed by Ulysses (e.g. Gosling 1996). More pointedly an important paper from Koyama et al. (1996) reports that *all class I sources in R Coronae Australis are hard X-ray, variable sources.* This information is to be added to the previous discovery by Montmerle et al. (1983) that the T Tauri stars (identified with classes II and III in the evolutionary sequence) emit soft X-rays prodigiously. It therefore seems clear that the central star is playing an important role in the general protostar activity, virtually as soon as it is mainly complete, although still in an accretion phase.

Koyama et al. (1996) also report flaring outbursts in the hard X-rays on a time scale of hours. If the twinned emission lines at 6.2 kev and 6.8 kev are an indication of bipolarity, then these stars must be ejecting material at velocities approaching 0.1c or some 30 times that of the fastest CMEs from the sun!

Of course there is the more usual evidence for short time variability as seen in molecular 'bullets', successive bow shocks, and general clumpiness in the jets.

Now flaring normally involves some particularly active subset of the global phenomenon and so I consider an unconventional local view of the infall/outflow. Consider a flux/vortex tube embedded in infalling material at a large distance from the central core. I suppose that the dissipation of electrical current and of vorticity occurs externally to the tube so that internally I can ignore 'resistivity' and 'viscosity'. In fact one ignores all internal forces so that this flux/vortex tube is conveniently thought of as being 'force-free'. Then there is a precisely parallel treatment of the internal magnetic field and vorticity that gives in the usual limits and notations

$$\frac{dB^i}{dt} = \sigma^{ij} B_j \, , \tag{1}$$

$$\frac{d\omega^i}{dt} = \sigma^{ij} \omega_j \, , \tag{2}$$

in differential form (the time derivatives are taken following the material). The shear or rate of strain tensor σ^{ij} is defined as

$$\sigma^{ij} \equiv \frac{1}{2}(\nabla^i v^j - \nabla^j v^i) \, . \tag{3}$$

By introducing the Lagrangian position of a fluid element $\vec{R}(t, \vec{r})$, where $\vec{r}$ is the initial position of the element, these equations integrate (so long as $| \, \partial \vec{R}/\partial \vec{r} \, |$ is not singular) to

$$\frac{\vec{B}}{\rho} = \left(\frac{\vec{B}_o}{\rho_o} . \vec{\nabla}_{\vec{r}} \right) \vec{R}(t, \vec{r}) \, , \tag{4}$$

$$\frac{\vec{\omega}}{\rho} = \left(\frac{\vec{\omega}_o}{\rho_o} . \vec{\nabla}_{\vec{r}} \right) \vec{R}(t, \vec{r}) \, . \tag{5}$$

Finally the density is given as usual by the conservation of mass as

$$\frac{d\ln\rho}{dt} = -g_{ij}\sigma^{ij} \equiv -Tr(\vec{\sigma}) \, . \tag{6}$$

Now I only wish to use these equations qualitatively here to encourage a new way of thinking about disc formation and flaring activity. Imagine first a flux/vortex tube that begins directed parallel to the rotation axis. In a purely radial collapse of the ambient cloud it is easy to see from these equations that the density, magnetic energy ($\vec{B} \cdot (1)$) and enstrophy ($\vec{\omega} \cdot (2)$) all increase. Moreover it is equally clear that the flux/vortex tube will become radial as it approaches the centre of the system.

However this neglects the collisions with other tubes in a possibly turbulent environment that can lead to reconnection and dissipation. Let us then suppose that the flux tube becomes incorporated into the central object, but that some undefined process of reconnection has allowed the tube to form a closed loop in the meridional plane. In fact we may imagine that there were initially an axi-symmetric distribution of flux/vortex tubes which now form an axi-symmetric 'arcade'. The outer surface of this arcade consists of interacting flux/vortex tubes that combine to form a vortex/current sheet.

Now I observe that such a vorticity distribution (i.e. poloidal) *implies* by the Biot-Savart law that adjacent material will be in rotation, at least near the equatorial plane. Moreover the current flow on the surface associated with the internally force-free flux tubes (mainly poloidal) implies an azimuthal magnetic field in the adjacent material. Thus it appears that *the system can not adopt such a reconnected configuration without implicating larger scale vorticity or rotation in the process.* The reconnection event then has been by way of vorticity/magnetic field shearing, probably in a disc-like structure in the equatorial plane. This is an archetypical flaring event, and may be the means by which excess magnetic flux and angular momentum is shed. This is similar to the earlier steady-state pictures referred to above, but differs in being essentially episodic as flux/vorticity tubes continue to reconnect. These may well be the events producing the X-ray sources and the optical jets (collimated by the previously evolved infall/outflow–lately known as the egg and the chicken).

Now one simple version of the possible transitional state may be calculated. Let us suppose that material falls in spherically on cones even though it is rotating. For additional simplicity we ignore the back reaction on the gas of magnetic field and centrifugal forces so that it is essentially freely falling and conserves angular momentum. Then from equations (4) and (5) above there follows for the dominant azimuthal component

$$\frac{B_\phi}{\rho} = \sin(\phi - \phi_o)\left(\frac{B_{o\theta}\cos\theta}{\rho_o} + \frac{B_{or}\sin\theta}{\rho_o}\right)$$
$$+ \frac{B_{or}\sin\theta}{\rho_o}\cos(\phi - \phi_o)r\partial_r(\phi - \phi_o(r)), \qquad (7)$$

and an exactly similar equation for the vorticity (replace B_ϕ by ω_ϕ). Here r is the initial radius of the gas, and $\phi_o(r)$ is its initial azimuth. This result holds provided that $\phi-\phi_o\lesssim\pi/2$ so that the transformation matrix remains strictly non-singular. It is the last term that dominates near the centre and we may calculate $\phi - \phi_o$ as approximately

$$\phi - \phi_o \approx \frac{\ell(r,\theta)}{\sin^2\theta}\sqrt{\left(\frac{2}{Gm(r)r}\right)}, \qquad (8)$$

for zero energy free-fall collapse along the cone. Here $\ell(r,\theta)$ is the specific angular momentum on each cone and $m(r)$ is the mass interior to the initial radius r. Notice that because of the dependence on B_{or} in the dominant term, the azimuthal flux/vortex tube will have the opposite sign above and below the equatorial plane.

If the axially symmetric distribution of flux tubes imagined above are thus transformed and incorporated into the surface of the star, then we have what one might refer to as the 'Jupiter effect' where azimuthally directed vortex and current sheets exist near the equatorial plane but with opposite senses above and below this plane. Such a situation is ripe for the reconnection inferred above, which is likely to be the flaring event. We note in passing that by the Biot-Savart law these current/vorticity sheets are associated near the plane with magnetic fields and velocities in the z direction of order $I/(2\Delta z)$ and $\Gamma/(2\Delta z)$ respectively. These are bipolar because of the sign reversal across the plane. Γ is the circulation on the sheets and I is the azimuthal current. Δz is a measure of their vertical extent. I conclude that such a configuration is associated with a bipolar outflow. Moreover the 'Jupiter effect' configuration is certainly unstable on short timescales. A fraction of the dissipating energy is likely to go directly into the induced bipolar outflow and so produce variable bipolar flaring.

Thus I have called attention to two time scales for variability. One is the evolutionary timescale as the central star becomes more active and for the massive stars at least develops a high velocity wind. The other is the inevitable flaring in the boundary layer which may exist throughout the evolutionary history, and which seems capable of creating high-energy bipolar ejections.

Acknowledgements

I am grateful to Silvano Massaglia and his organizing committee for inviting me to this city of 'arcades'. The work was supported in part by the National Sciences and Engineering Research Council of Canada.

REFERENCES

Bachiller, R., 1996, *Ann. Rev. Astron. Astrophys.*, **34**, 111.

Blandford, R.D., & Payne, D.G., 1982, *MNRAS*, **199**, 883.

Bontemps, S., André, P., Tereby, S., & Cabrit, S., 1996, *A&A*, **311**, 858.

Chen, X., Taam, R.E., Abramowicz, M.A., & Igumenschev, I.V., 1997, *MNRAS*, **285**, 439.

Clarke, D. Karpik, S., & Henriksen, R.N., 1985, *Ap.J.S.*, **58**, 81.

Ferreira, J., & Pelletier, G., 1995, *A&A*, **295**, 807.

Fiege, J., & Henriksen, R.N., 1996a,b, *MNRAS*, **281**, 1038 and 1055.

Gosling, J.T., 1996, *Ann. Rev. Astron. Astrophys.*, **34**, 35.

Henriksen, R.N., 1987, *Ap.J.*, **314**, 33.

Henriksen, R.N., & Valls-Gabaud, D., 1994, *MNRAS*, **266**, 681.

Henriksen, R.N., André, P., & Bontemps, S., 1997, *A&A*, in press.

Koyama, K., Hamaguchi, K., Ueno, S., Kobayashi, N., & Feigelson, E.D., 1996, *PASJ*, **48**, L87.

Masson, C.R., & Chernin, L.M., 1993, *Ap.J.*, **414**, 230.

Montmerle, Th., Koch-Miramond, L., Falgarone, E., & Grindlay, J.E., 1983, *Ap.J.*, **269**, 182.

Najita, J.R., 1995, *Rev. Mex. Astron. Astrof. (ser. conf.)*, 1, 293.

Ouyed, R., Pudritz, R.E., & Stone, J.M., 1997, *Nature*, **385**, 409.

Palla, F., & Stahler, S.W., 1990, *Ap.J.*, **360**, L47.

Raga, A., & Cabrit, S., 1993, *A&A*, **278**, 267.

Toropin, Yu.M., Saveljev, V.V., & Chechetkin, V.M., 1996, preprint at toropin@sai.msu.su.

Tsinganos, K., Sauty, C., Surlantzis, G., Trussoni, E., & Contopoulos, J., 1996, *MNRAS*, in press.

CHAPTER 18

THE IMPLICATIONS OF OBSERVATIONS ON MODELS OF HERBIG-HARO OBJECTS

A.C. RAGA AND J. CANTÓ

Instituto de Astronomía, UNAM, Ap. 70-264, 04510 México D. F., México

A few Herbig-Haro (HH) objects have now been observed with the Hubble Space Telescope (HST). These observations have clearly resolved the cooling regions behind the shocks associated with these objects for the first time. Problems arise in trying to model the HST observations, as they show a separation between the bow shock and Mach disc of the same order as the jet diameter. We propose a model in which the large bow shock/Mach disc separation results from the presence a cool "nose cone" produced by a non-top hat ram-pressure cross section of the jet. The conditions necessary for producing a "magnetic nose cone" are also discussed, though this possibility is not pursued in detail in the present paper.

1. INTRODUCTION

One of the longstanding problems in observations of HH objects has been the lack of a clear spatial resolution of the recombination regions behind the shock waves associated with these objects. This has seriously limited the possible comparisons between shock models and observations.

In a very interesting paper, Schwartz (1981) noted that HH emission lines of different excitation/ionization had different radial velocities. He suggested that this might be an indirect indication of the postshock ionization stratification. A second indirect observation of the postshock stratification of HH objects was carried out by Solf, Böhm and Raga (1988). These authors obtained temperature diagnostics from lines of many ions (e. g., from [O I], [O II] and [O III] lines), and found a monotonic growth of electron temperature with ionization potential of the parent ion.

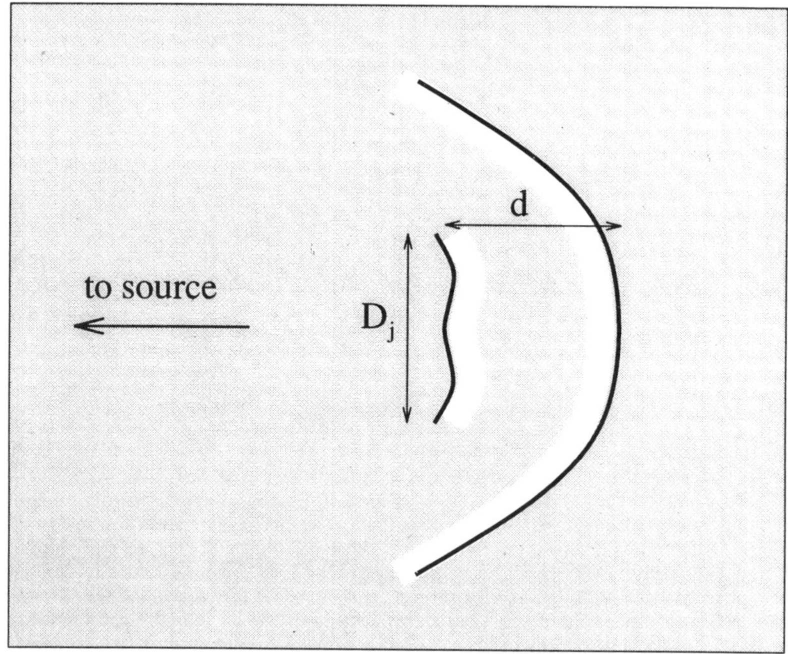

FIGURE 1. This figure shows a schematic diagram of the HST observations of HH 47A of Heathcote et al. (1996). Two high intensity gradient regions delineate what appear to be the bow shock and Mach disc of a working surface. The immediate post shock regions are observed to be bright in Hα (regions shaded in white), and the region in between the two shocks is filled with a diffuse emission which is bright in the red [S II] lines.

It is of course more interesting to obtain a direct spatial resolution of the recombination regions of HH shock waves. Small ($\sim 0.5''$) offsets between the Hα and [S II] 6717+31 emission have been detected for the HH 34S (Bührke, Mundt and Ray 1988) and HH 111 V (Reipurth, Raga and Heathcote 1992) bow shocks. Raga and Binette (1991) showed that these offsets were consistent with the ones expected from one-dimensional, stationary shock models of velocities $\sim 10\text{-}80$ km s^{-1}.

Much more impressive are the results obtained by Heathcote et al. (1996) for HH 47A, the brightest working surface of the HH 46/47 outflow. Figure 1 is a schematic diagram of the results obtained by these authors. Two regions of high intensity gradient appear to delineate the loci of the bow shock and Mach disc of HH 47A. The immediate post shock regions are brighter in Hα than in the [S II] 6717+31 emission. These two "Hα ridges" are schematically shown in Fig. 1. The extended region in between the two shocks has a diffuse emission with bright [S II] and weak Hα. These results apparently are a clear resolution of the recombination regions behind the Mach disc and bow shock of HH 47A.

It is notable that both the diameter of the Mach disc D_j and the separation between Mach disc and bow shock d (see Fig. 1) have angular sizes of $\sim 2\text{-}3''$, corresponding to a physical size of $\sim 1.5 \times 10^{16}$ cm. As we will see in the following section, this result is quite surprising, since from radiative working surface models we would expect to have $d \ll D_j$. This point is discussed in §2, and possible ways of explaining the $d \sim D_j$ observational result are discussed in §§3 and 4.

2. THE BOW SHOCK/MACH DISC SEPARATION

Let us carry out a simple estimate of the separation between the Mach disc and bow shock of a radiative working surface. Standing in a reference system that moves with the working surface, we have a situation such as shown in Fig. 2.

Using the usual mass conservation arguments and the strong shock jump conditions, one can straightforwardly show that the separations between the Mach disc, contact discontinuity and bow shock d_1, d_2 and d (see Fig. 2) are given by:

$$d = d_1 + d_2 \approx \frac{(1+\beta)^2}{2\beta} \frac{r_j}{M_j{}^0} \; ; \qquad \frac{d_1}{d_2} \approx \beta \, , \tag{1}$$

where $M_j{}^0 = v_j/c_0$ is the jet Mach number with respect to the sound speed of the working surface material, $\beta = \sqrt{\rho_j/\rho_e}$, and r_j is the radius of the jet close to the working surface. For deriving Eq. (1), one also has to assume that the material in between the two shocks has a uniform sound speed c_0 (see the caption of Fig. 2).

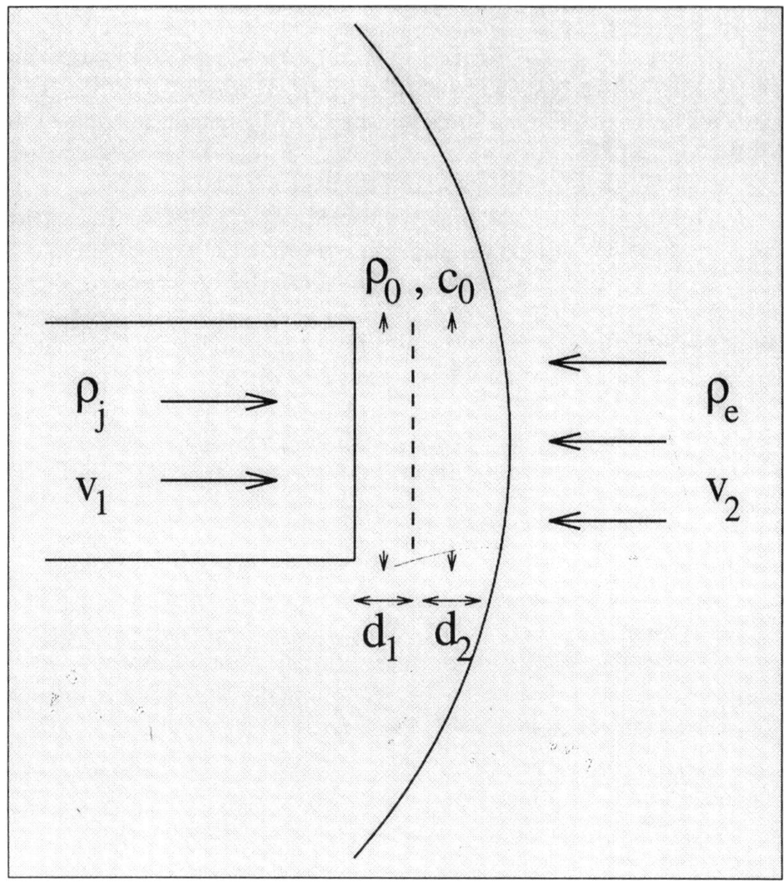

FIGURE 2. Schematic diagram of a radiative working surface, seen from a system at rest relative to the shock structure. Material from the jet (of density ρ_j) is fed into the working surface at a velocity v_1. The environment (of density ρ_e) enters the bow shock with a velocity v_2. The material in between the two shocks immediately cools to a temperature of $\sim 10^4$ K, so that it has an approximately uniform sound speed c_0 (and density ρ_0). This material exits the working surface laterally with a velocity $\sim c_0$. The contact discontinuity (dashed line) that separates the jet and environmental material is located at a distance d_1 from the Mach disc, and a distance d_2 from the bow shock. The Mach disc-bow shock separation is $d = d_1 + d_2$.

From these arguments, we see that for an HH jet with $M_j{}^0 \sim 10$-30 and $\beta \sim 1$-10, the separation between the working surface shocks has a value $d \sim 0.07$-$0.6\,r_j$ (see Eq. 1). This appears to be in clear disagreement with the HST observations of HH 47A (see Heathcote et al. 1996 and Fig. 1) which appear to imply a separation of $d \approx D_j = 2\,r_j$.

Hα - [S II]   t = 300 yr

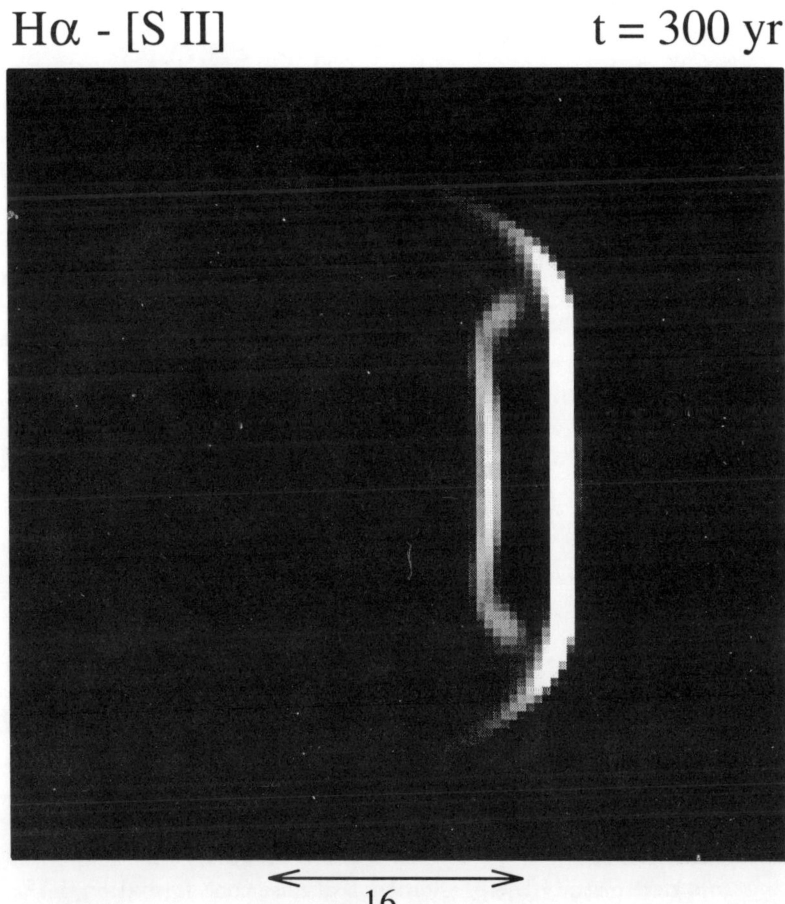

FIGURE 3. Subtraction of the [S II] intensity map from the Hα map predicted from the top-hat jet simulation described in the text. The positive regions (with Hα brighter than [S II]) are in white, and the darker colors indicate the negative ([S II] bright) regions.

We find that Eq. (1) reproduces quite well the results obtained from axisymmetric numerical simulations of radiative working surfaces. We have carried out a simulation of a top-hat cross section jet of radius $r_j = 10^{16}$ cm, velocity $v_j = 120$ km s^{-1}, density $n_j = 60$ cm^{-3}

and temperature $T_j = 10^4$ K travelling into an environment of density $n_e = 10$ cm^{-3} and temperature $T_j = 10$ K. The working surface is maintained inside the computational grid by dividing v_j into velocities v_1 and v_2 (see Fig. 2) such that the jet/environment ram pressure balance condition is met.

Figure 3 shows some results from this simulation. In particular, this figure shows a subtraction of the predicted [S II] intensity map from the Hα map. This corresponds to the technique used by Heathcote et al. (1996) for illustrating the Hα emission ridges delineating the bow shock and Mach disc of HH 47A.

As is clear from Fig. 3, we do get such Hα ridges, and we also obtain the extended [S II] emission in the region between the Mach disc and bow shock. However, we get a small separation between the Mach disc and bow shock, which is in relatively good agreement (within a factor 2) of the value predicted from Eq. (1). This is of course an important problem in trying to interpret the observations of HH 47A with a working surface model, and we will address it in the following section.

3. HYDRODYNAMIC NOSE CONES

It has been noted, e. g. by Blondin, Fryxell and Königl (1990), that numerical simulations of radiative jets sometimes develop a plug of cool gas that piles up between the Mach disc and the bow shock of the working surface. Such so called "nose cones" are the result of a non-top hat cross section for the ram pressure of the jet beam. Such cross sections are sometimes attained in numerical simulations of "long jets" as a result of the numerical diffusion of the integration.

In order to study the phenomenon of nose cone formation, let us consider the ram pressure balance solution for a working surface of a jet with an arbitrary ram pressure cross section $\rho_j\, v_j{}^2(r)$ moving into a uniform environment. If we stand in a reference system such that we satisfy on-axis jet/environment ram pressure balance, we have a situation such as the one shown in Fig. 4.

As the ram pressure of the environment (see Fig. 4) is uniform, if we have on-axis ram pressure balance, this condition will clearly not be satisfied off-axis. Because of this, the off-axis region of the working surface develops a curved structure which can be defined with the offset $x(r, t)$ (see Fig. 4). From the usual ram-pressure balance argument, it can be straightforwardly shown that this offset satisfies the equation:

$$x(r,t) = \left(v_e - \sqrt{\frac{\rho_j \, v_j{}^2(r)}{\rho_e}} \right) t \, ; \quad v_e \equiv \sqrt{\frac{\rho_j \, v_j{}^2(0)}{\rho_e}} \, . \qquad (2)$$

For example, if the ram pressure of the jet peaks on-axis, Eq. (2) predicts that the working surface will curve more and more towards the source with increasing time (this is the case illustrated in Fig. 4).

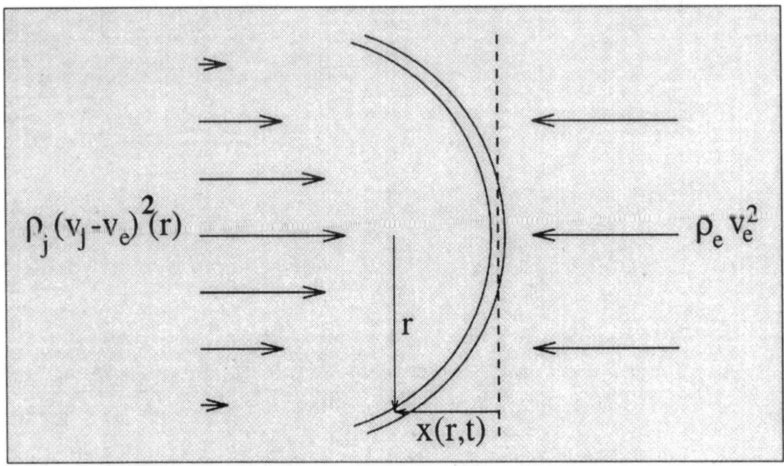

FIGURE 4. Working surface produced at the head of a jet with a non-top hat ram pressure cross section. If we stand in a reference system at rest with the tip of the working surface, we find that the off-axis regions of the working surface curve away more and more with increasing time.

Such a curved working surface results in an interesting flow geometry: while the post-bow shock material flows away from the symmetry axis, the post-Mach disc material flows towards the axis. This shocked jet gas then starts to pile up in the stagnation region of the working surface, forcing apart the Mach disc and the bow shock. In this way, a "nose cone" of cool, shocked jet gas starts to form.

This can be seen in the results from a numerical simulation with parameters identical to the ones of the model of §2, but with a Gaussian density cross section (with a centre-to-limb density ratio of 6). An Hα-[S II] image resulting from this simulation is shown in Fig. 5, where we can see that the separation d between the Mach disc and the bow shock does indeed become comparable to the diameter of the Mach disc.

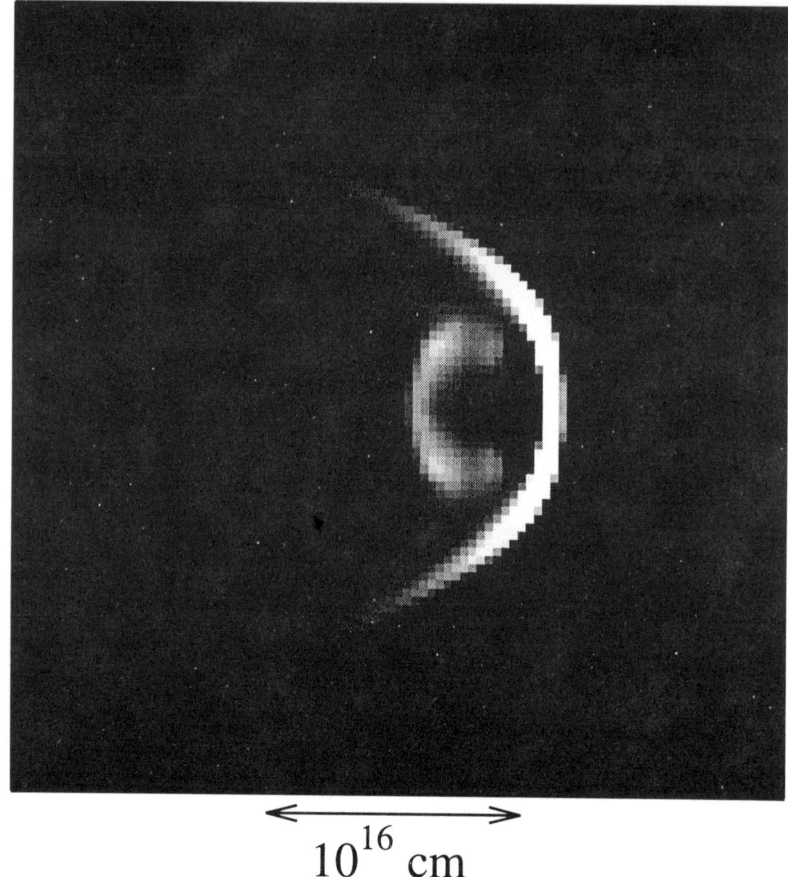

FIGURE 5. Subtraction of the [S II] intensity map from the Hα map predicted from the variable ram pressure jet simulation described in the text. The positive regions (with Hα brighter than [S II]) are in white, and the darker colors indicate the negative ([S II] bright) regions.

The image shown in Fig. 5 does show a strong qualitative resemblance to the image of HH 47A of Heathcote et al. (1996), which is shown schematically in Fig. 1. From this we conclude that the large, diffuse [S II] bright region separating the Mach disc and bow shock of HH 47A might correspond to a gasdynamic nose cone formed as a result of a variable ram-pressure cross section of the jet.

3. OTHER POSSIBILITIES

The clear second possibility for explaining the HST observations of HH 47A is that the large Mach disc/bow shock separation is due to the effects of a strong magnetic field. MHD simulations of extragalactic jets sometimes show the formation of "magnetic nose cones" (see, e. g., Clarke 1990; Kössl, Muller and Hillebrandt 1990). From simple arguments, it is possible to show that in order to confine the gas between the two working surface shocks the magnetic field B in this region has to satisfy the condition :

$$B \geq 740\mu G \left(\frac{\beta}{1+\beta} \right) \left(\frac{n_e}{100\,\mathrm{cm}^{-3}} \right)^{1/2} \left(\frac{v_j}{100\,\mathrm{km\,s}^{-1}} \right), \qquad (3)$$

where n_e is the electron density of the nose cone and $\beta = \sqrt{\rho_j/\rho_e}$. Of course, the magnetic field also has to have an appropriate topology to be able to confine the working surface material.

Even though the magnetic field resulting from Eq. (3) appears to be quite large, the possibility of the existence of magnetic nose cones in HH jets cannot be discarded completely, and deserves further study.

Acknowledgements

The authors acknowledge support from CONACYT, DGAPA (UNAM) and the UNAM-Cray program.

REFERENCES

Blondin, J.M., Fryxell, B.A., & Königl, A., 1990, *Ap.J.*, **360**, 370.

Bührke, T., Mundt, R., & Ray, T.P., 1988, *A&A*, **200**, 99.

Clarke, D.A., 1990, in *Galactic and Extragalactic Magnetic Fields*, IAU Symp. 140, eds. R. Beck, P. P. Kronberg and R. Wielebinski (Reidel: Dordrecht), p. 403.

Heathcote, S., Morse, J.A., Hartigan, P., Reipurth, B., Schwartz, R.D., Bally, J., & Stone, J.M., 1996, *A.J.*, **112**, 1141.

Kössl, D., Muller, E., & Hillebrandt, W., 1990, *A&A*, **229**, 397.

Raga, A.C., & Binette, L., 1991, *Rev.Mex.AA*, **22**, 265.

Reipurth, B., Raga, A.C., & Heathcote, S., 1992, *Ap.J.*, **392**, 145.

Schwartz, R.D., 1981, *Ap.J.*, **243**, 197.

Solf, J., Böhm, K.H., & Raga, A.C., 1988, *Ap.J.*, **334**, 229.

CHAPTER 19

THE PRECESSING JETS OF SS433

W. BRINKMANN[1] AND E. MÜLLER[2]

[1] Max-Planck-Institut für extraterrestrische Physik,
85740 Garching, Germany

[2] Max-Planck-Institut für Astrophysik, 85740 Garching,
Germany

The precessing jets of SS433 are modeled with a time dependent 3-dimensional hydro code. As they are the observationally by far best studied astrophysical jets and all their relevant physical parameters are known with high accuracy, they present an ideal laboratory to confront numerical jet simulations with reality. Although our simulations have not yet reached the evolutionary age of the 'real' jets, most of the unexpected recent observational results can be explained quite naturally.

1. INTRODUCTION

The enigmatic Galactic object SS433 (Margon 1980) shows radio to X-ray emission from two oppositely directed jets with velocities of $v \sim$ 0.26 c. The compact object emitting the jets is member of a binary system with an orbital period of $P_0 = 13.1$ days and the jets follow a precessional motion with a $\sim$ 162.5 day period with a precession angle of $\theta = 19.80°$. This picture is confirmed in a vast number of observations which allowed the accurate determination of the physical parameters of the 'optical' jet, consisting of cold material ($T \lesssim 10^4 \mathrm{K}$)

emitting the optical light at distances 10^{14} cm $\leq r_{opt} \leq 10^{15}$ cm from the jet's origin (for a review see Margon 1984). VLA observations confirmed the distance of the object of 5 kpc and yielded further direct evidence for the precessional motion of the matter of the jets (Hjellming and Johnston 1981). Further, SS433 is at the center of the supernova remnant W50 which has an estimated age of $\sim (2-5) \times 10^4$ years.

X-ray observations with the EXOSAT satellite revealed Doppler shifted iron lines in the spectrum of SS433 (Watson et al. 1986) with the line centroids following the precessional Doppler curve. In a series of observations with Ginga from 1987 to 1991, utilizing the LAC's high sensitivity and wide energy range (Kawai et al. 1989, Brinkmann et al. 1989, Brinkmann et al. 1991) and more recently with ASCA (Yamauchi et al. 1994, Kotani et al. 1996) the picture of thermally radiating jet was confirmed. The emission lines are superimposed on a very hot $(kT \gtrsim 20 \, keV)$ thermal continuum and vary with precessional phase, i.e., with the viewing angle. By hydrodynamical modeling the material parameters of the inner X-ray jets, like the temperature and density, as well as the geometrical dimensions of the jets could be constrained to within a factor of unity.

The general picture emerging from these multi-wavelength observations is as follows: the X-ray jet consists of hot 'normal' material at initial densities of $\rho_i \gtrsim 10^{13}$ cm^{-3}, flowing out at the final velocity of $\sim 0.26c$. The jets are hot, $T \gtrsim 20$ keV, and cool within a distance of $\sim 10^{12}$ cm to temperatures below 10^5 K (see Brinkmann et al. 1991). Further out, at distances of $\sim 10^{15}$ cm the jet emits the optical 'moving' lines, presumably from blobs of matter arising from thermal instabilities of the cooling X-ray jet (Bodo et al. 1988, Brinkmann et al. 1988). At distances of $10^{15} - 10^{17}$ cm the jets show strong radio emission $(\gtrsim 0.5$ Jy at 5 GHz with spectral index α=-0.6; c.f. Vermeulen 1993, for a review).

The jets then remain 'invisible' until the matter flow interacts with the circumstellar medium and disposes its energy and momentum. The jets first show up in X-rays again ~ 20 arc min from SS433, i.e., at distances $\gtrsim 1.5 \times 10^{18}$ cm. The radio shape of the host supernova remnant W50 shows two well known lobes, the so-called 'ears', caused by the out-flowing jets which seem to be confined to the shell of W50. In the radio band the out-flowing matter can be seen only very close to SS433, on scales of 100 milli arcsec (Hjellming and Johnston 1981) and further out, at the 'ears'. In between, there are no radio contours indicating the presence of a directed, collimated flow.

However, X-ray observations with imaging X-ray instruments demonstrated that energy is continuously supplied to the surrounding. Observations with the high spatial and moderate spectral resolution of the ROSAT PSPC (0.1 - 2.4 keV) allowed the study of spectral variations along and across the jets as well as of the spatial structure of the jets (Brinkmann et al. 1996). The emission appears patchy, peaks at distances of $\sim$ 35 arc min from the source and shows hardly any flux close to SS433. The main contribution to the flux comes from the narrow cone of the jets of about 20 degrees opening angle, but there is, at a lower level, a much wider component suggesting that energy is 'leaking out' from the well confined jets. The soft X-ray luminosity from each lobe is $\sim 6 \times 10^{34}$ ergs s^{-1}. It turned out that the X-ray emission from the two jets differs considerably, morphologically and spectrally. It is not expected that the different Doppler boosting of the jets can account for these asymmetries, nor that the jets are intrinsically very different. The conclusion is that the breaking of the jets and, correspondingly, its high energy emission, critically depends on the boundary conditions imposed by the surrounding medium.

ASCA observations of the eastern jet show that the spectrum can be best fitted with a power law with photon index in the range $\Gamma \sim 1.4$ - 2.2 (Yamauchi et al. 1994), i.e., the emission appears to be non-thermal.

The observational results thus pose some questions which are hard to understand in the 'standard kinematical jet model':

1. ROSAT observations of the SS433/W50 system clearly show that there is a large mismatch between the energy input from the jets into the nebula and the total emission from W50 by a factor of $> 10^4$. It remains unclear where this excess energy is deposited.

2. The extent and shape of the X-ray jets is not as expected from the 'canonical' precessing jet model. Whether a 'refocusing' of the precessing jets (Peter and Eichler 1993) can account for this remains to be studied.

3. ASCA and ROSAT observations indicate that most of the extended jet's X-ray emission is non-thermal. The conversion of the jet's kinetic energy into relativistic particles is not yet understood.

4. The X-ray emission anti-correlates with the radio emission (at least for the eastern jet) - the maximum of the X-ray emission coincides spatially with a 'hole' in the radio maps.

These questions can only be answered by a detailed modeling of the jet propagation and its interaction with the surrounding ISM. So far, the main difficulty inhibiting all attempts for a 'proper' treatment was that these jets are intrinsically 3 dimensional, requiring large CPU time and memory resources.

However, as all relevant parameters of the jets are known to high

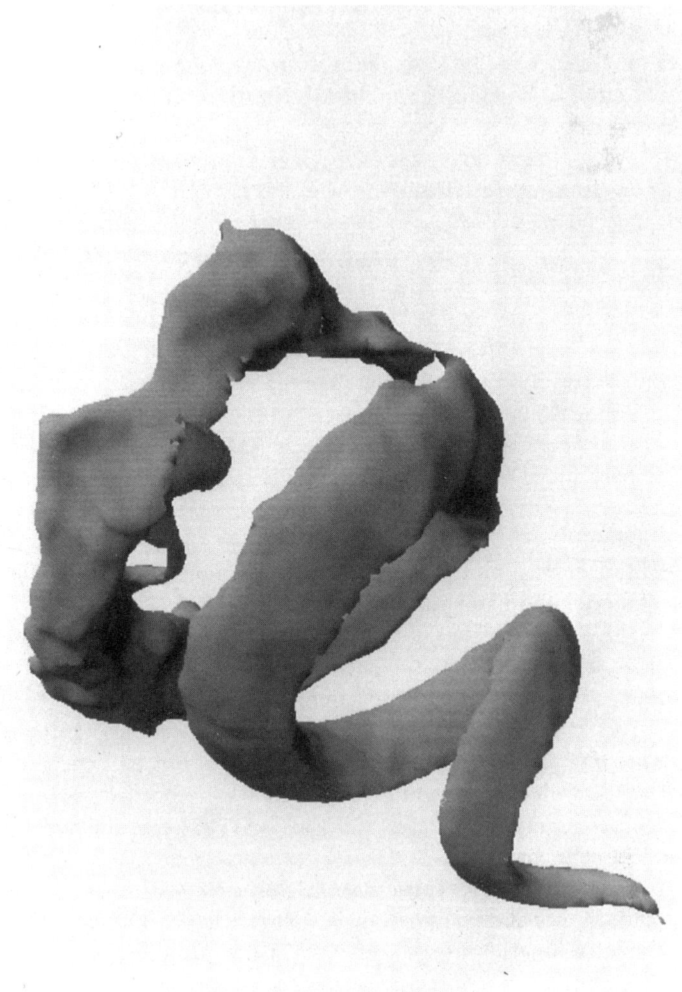

FIGURE 1. The three dimensional surface of the SS433 jet containing more than 60% of the original jet material after six precession periods (see Color Plate 4).

accuracy the jets of SS433 present the currently best laboratory for the study of the jet phenomenon in general and thus for confronting numerical jet models with observations.

2. THE NUMERICAL CALCULATIONS

We study the hydrodynamic temporal evolution of the precessing jets with the explicit 3-D Eulerian PPM hydro-code **Prometheus** (Fryxell et al. 1989). We employed a spherical coordinate system with $80 \times 80 \times 500$ zones, corresponding to an angular region of $80^\circ \times 80^\circ$ and a non-uniformly spaced radial grid from 10^{16} cm to 5×10^{17} cm.

At the outer boundaries we used outflow boundary conditions. At the inner (radial) boundary the time dependent precessing jet is flowing into the computational grid. As flow parameters of the jet typical values determined from the observations as discussed above were taken. Up to a distance corresponding to the innermost grid zone the jet can be regarded as straight and conical. Only well into our computational grid a kinematically out-flowing jet shows the typical cork-screw pattern due to retardation effects.

The background matter was taken to be cold ($T \sim 10^4$ K), uniform hydrogen at a density of 1 cm^{-3}. Due to the strong interaction with the Mach ~ 800 flow the shocked gas reaches temperatures well in excess of 10^9K so that a relativistic equation of state had to be used.

The computations were done on a 16 processor Cray Jedi at the RZG Garching. One day in the 'real' life of SS433 typically takes ~ 1 hour of CPU time. So far we have calculated eight precessional periods of the jet. One of the main technical problems is, in fact, the graphical presentation of the data. As expected, we find very complex interactions of the successive 'windings' of the outstreaming cork screw pattern with the ISM and with the flow pattern generated by the previous precession periods. These structures in a conically expanding flow pattern are really 3-dimensional and hard to visualize in 2-D. Details can be found in Müller and Brinkmann (1997). Recently, we are performing a second simulation ($500 \times 120 \times 120$ zones) where the angular size of the computational volume is enlarged ($170^\circ \times 170^\circ$) to reduce unwanted boundary effects.

Figure 1 shows a jet after it evolved for six precessional periods. Plotted is the three dimensional surface of the computational volumes which contain more than 60% of the original jet material. Close to the origin the jet has its undisturbed original shape, but fairly soon it gets heavily deformed through its strong, complex interaction with the circumstellar medium. Note that only about four of the precession periods can be clearly distinguished, the material in the front of the jet has been swept up into an irregular structure due to the radial velocity gradients evolving in the flow.

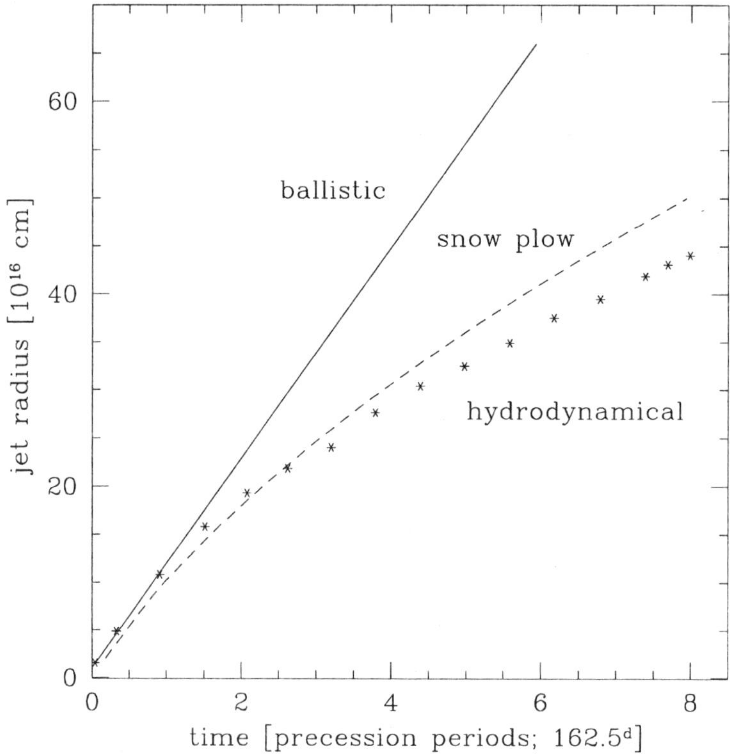

FIGURE 2. Radial advancement of the head of the jet as function of time. Full line: ballistic outflow; stars: results of our computations; dashed line: snowplow model.

3. FIRST PHYSICAL INTERPRETATIONS

As mentioned above the head of the out-flowing jet gets slowed down by sweeping up the ISM in front of it and merges with matter from subsequent periods of the jet. Thus the bulk radial advancement of the jet does not take place at the jets initial velocity but at a much lower velocity.

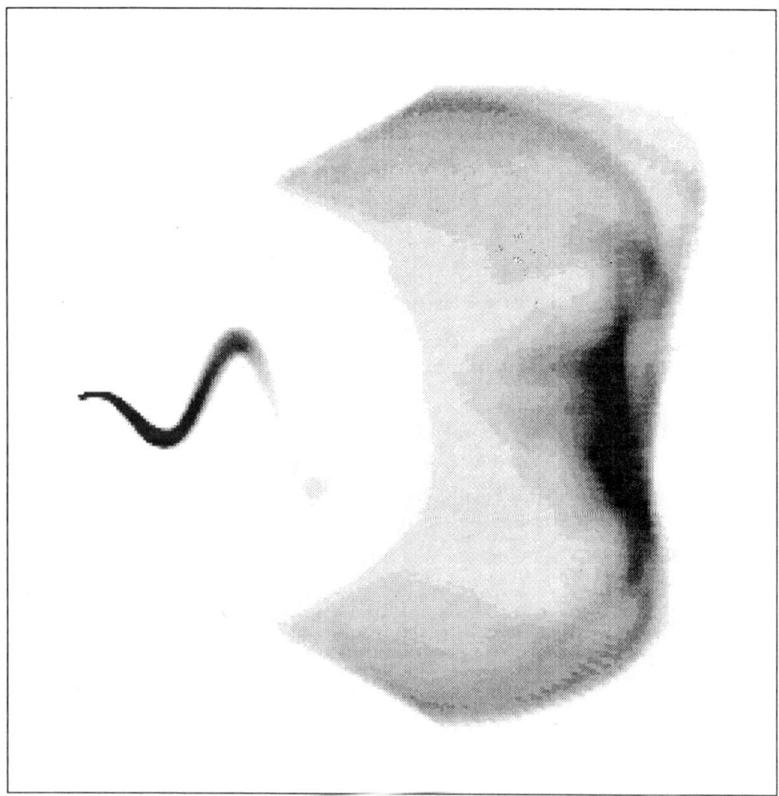

FIGURE 3. Spatial distribution of the soft X-ray (0.1 - 2.4 keV) from the simulated jet. For details see text.

This speed can be determined approximately by momentum balance of the matter at the head of the jet as it moves outwards, engulfing the ISM in front (snowplow effect). In Fig. 2 we compare the predictions of a purely ballistic outflow (with 0.26 c), to the results obtained from our numerical computations, and the predictions of the snowplow model. As can be seen, the snowplow model represents a surprisingly good approximation to the complex 3-dimensional situation.

Extrapolating this approximation to the current dimension of the SS433/W50 system we get a dynamical estimate of the age of W50, depending on the unknown value of the initial density of the ISM. For the parameters used in our calculation the age of W50 would be greater than 10^5 years, i.e., at least a factor of two more than currently thought.

Although the numerical simulations have by far not reached the evolutionary age of the SS433/W50 system we can ask whether there are indications for a solution of the two main questions, raised by the X-ray observations: 1.) what is the origin of the apparently featureless power law spectrum detected by ASCA and ROSAT and, 2.) why is the X-ray emission seen by ROSAT concentrated in a much smaller angular region than expected from the $\sim 40^o$ opening angle of the precessing jets?

We calculated the X-ray emission from the numerical model by summing up the thermal emission from each of the 3.2×10^6 volume elements. For each volume we determined the equilibrium ionization structure at the given temperature, assuming cosmic matter composition (Allen 1973). The ionization and recombination rates were taken from Arnaud and Rothenflug (1983), with corrections for the Fe rates from Arnaud and Raymond (1992). For the calculation of the optically thin X-ray emission we employed the emission model of Mewe et al. (1985).

It turns out that, due to the high temperatures obtained in the jet - ISM interaction, the integrated X-ray spectrum is relatively featureless. The iron line has an equivalent width of $\lesssim 20$ eV and might thus have been missed by previous X-ray detectors. Only future, more sensitive instruments will be able to distinguish this kind of spectrum from a truly non-thermal power law. In Fig. 3 we show the corresponding X-ray emission as it would be seen with an imaging X-ray detector in the ROSAT (0.1 - 2.4 keV) energy band, under the appropriate geometrical viewing conditions for the SS433/W50 system. Clearly visible are the boundaries of our computational domain. Please note that the bulk of the emission does not originate from the jets themselves, but from the interior region closer to the axis. The opening angle is thus not the full precessional cone, but it is considerably smaller, as seen in the ROSAT measurements.

4. OUTLOOK

We have presented some first results of the hydrodynamic simulations of the initial phase of the evolution of the precessing jets of SS433. Although the computer simulations have not been performed up to the evolutionary age of the real jets, it is nevertheless indicated that the numerically deduced physical properties seem to be able to explain the observational results quite naturally.

REFERENCES

Allen, C.W., 1973, Astrophysical Quantities, 3rd ed., Athlone Press, London

Arnaud, M., & Rothenflug, R., 1985, *A&AS*, **60**, 425.

Arnaud, M., & Raymond, J., 1992, *Ap.J.*, **398**, 394.

Bodo, G., Ferrari, A., Massaglia, S., & Brinkmann, W., 1988, *Astrophys. Lett. Comm.*, **27**, 5.

Brinkmann, W., Fink, H.H., Massaglia, S., Bodo, G., & Ferrari, A., 1988, *A&A*, **196**, 313.

Brinkmann, W., Kawai, N., & Matsuoka, M., 1989, *A&A*, **218**, L13.

Brinkmann, W., Kawai, N., Matsuoka, M., & Fink, H.H., 1991, *A&A*, **241**, 112.

Brinkmann, W., Aschenbach, B., & Kawai, N., 1996, *A&A*, **312**, 306.

Fryxell, B., Müller, E., & Arnet, D., 1989, MPA preprint 449.

Hjellming, R.M., & Johnston, K.J., 1981, *Ap.J.*, **246**, L141.

Kawai, N., Matsuoka, M., Pan, II.C., & Stewart, G.C., 1989, *PASP*, **41**, 491.

Kotani, T., Kawai, N., Matsuoka, M., & Brinkmann, W., 1996, *PASJ*, **48**, 619.

Margon, B., 1980, *Ann. N.Y. Acad. Sci.*, **336**, 550.

Margon, B., 1984, *Ann. Rev. Astr. Astr.*, **22**, 507.

Mewe, R., Gronenschild, E.H.B.M., & Van den Oord, G.H.J., 1985, *A&AS*, **62**, 197.

Müller, E., & Brinkmann, W., 1997, submitted to *A&A*

Peter, W., & Eichler, D., 1993, *Ap.J.*, **417**, 170.

Vermeulen, R., 1993, in: Astrophysical Jets, eds. R. Burgarella, M. Livio, C. O'Dea, Cambridge Univ. Press, Cambridge, p. 241.

Watson, M.G., Stewart, G.C., Brinkmann, W., & King, A.R., 1986, *MNRAS*, **222**, 261.

Yamauchi, S., Kawai, N., & Aoki, T., 1994, *PASJ*, **46**, L109.

CHAPTER 20

STABILITY OF RADIATIVE HERBIG-HARO JETS

S. MASSAGLIA

Dipartimento di Fisica Generale dell'Università, Via P. Giuria 1, I-10125 Torino, Italy

In this chapter I study the evolution of Kelvin-Helmholtz instabilities of supersonic jets in a 2-D cylindrical geometry in presence of radiative losses and heating. In particular, I examine the effects of different assumptions for the heating term that, initially, balances losses in the energy equation, namely: i) constant heating, ii) heating proportional to the particle density, and i) no heating. I find that the different choices, albeit affecting the details of the instability evolution, do not qualitatively change the general behavior.

1. INTRODUCTION

Herbig-Haro jets propagate, for long distances and at high Mach numbers, into the ambient of star formation regions and are subject to the potentially disruptive Kelvin-Helmholtz (K-H) instability. Nevertheless they are observed to reach, and sometimes exceed, the parsec scale (see Bally and Devine 1994, Ray Chapter 16). This implies that the various K-H modes, that develop in the jet while opening its way into the cloud's medium, are not effective enough to smear the jet out. Bodo et al. (1994, 1995) have shown that 2-D jets in adiabatic conditions may indeed undergo relevant mixing processes as due to the nonlinear evolution of K-H modes. Moreover, 3-D calculations

(Bodo, Chapter 15) point out that strong mixing develops earlier in this case than in the corresponding 2-D simulations, due to the growth of asymmetric modes.

Recent 2-D numerical calculations by Massaglia et al. (1996, MRBF) and Rossi et al. (1997, RBMF) have shown that radiative losses can be quite effective in reducing the mixing induced by Kelvin-Helmholtz instabilities. A problem one finds when including inhomogeneous terms in the energy conservation equation is the role played by local heating, that is hardly an observable quantity. One can only argue about the relative importance of the various physical processes (ionizing radiation from the central star, turbulence, resistive dissipation of magnetic fields, etc.) that can possibly contribute to the energy input. On the other hand the instability evolution may depend crucially on the form assumed by heating and its dependence on the physical variables.

Stone, Xu and Hardee (1997) studied the 'spatial' evolution (see below) of antisymmetric modes in a slab jet in presence of radiation losses and heating proportional to the particle density, and balancing the losses at equilibrium (see also Stone 1997). Other authors (e.g. Blondin, Fryxell and Königl 1990, Raga et al. 1995), in the context of Herbig-Haro jets propagation, considered purely radiative flows, without heating. In the 'temporal' analysis of MRBF and RBMF the heating term was kept constant and balancing losses at equilibrium. Massaglia et al. (1997) have tackled the same problem for a symmetric modes in a cylindrical jet, in the context of a 'spatial' analysis, considering the effects of constant heating and heating proportional to density. They found the two assumption leading to qualitatively identical instability evolution.

In this contribution I discuss how the above different choices for the heating term affect the instability evolution. I follow a 'temporal' approach, i.e. study the growth of the perturbations in time, as opposed to the 'spatial' approach, where the evolution of the modes is followed in space (e.g. Stone, Xu and Hardee (1997), Massaglia et al. 1997). As discussed in MRBF and RBMF, this approach has the advantage of allowing to study the instability over long evolutionary times.

2. THE PHYSICAL PROBLEM

The hydrodynamical equations of mass, momentum and energy conservation for a fluid jet, in 2-D cylindrical coordinates (r, z), have been solved numerically using a PPM code (Colella and Woodward

1984), modified to include radiative losses and heating in the energy equation by means of an operator splitting technique as described in MRBF and RBMF. I will also refer to MRBF and RBMF for the details of the radiation treatment and the characteristics of the initial configuration, perturbation, boundary conditions and the integration domain.

The energy equation can be written

$$\frac{\partial E}{\partial t} + \nabla \cdot (E\vec{v}) = -p(\nabla \cdot \vec{v}) + (\mathcal{H} - \mathcal{L}),$$

where the fluid variables p, $\vec{v}$ and E are pressure, velocity, and thermal energy per unit volume respectively; $\mathcal{H}$ and $\mathcal{L}$ represent the energy input and loss term (energy per unit volume per unit time), the latter including energy lost in line emission and in the ionization and recombination processes. In the radiative losses, we consider line emission from nine elements: H and He resonance lines, and the 13 strongest forbidden lines of C, N, O, S, Si, Fe and Mg, whose abundances have been assumed to be solar. The ionization state of H is obtained by integrating, along with the fluid equations the temporal evolution equation for the number fraction of neutral hydrogen atoms, f_n:

$$\frac{\partial f_n}{\partial t} + (\vec{v} \cdot \nabla)f_n = n_e \left[-(c_i + c_r)f_n + c_r \right],$$

which includes collisional ionization and recombination, and where n_e is the electron density, c_i and c_r are the ionization and recombination rate coefficients (see RBMF). As for the ionization state of the other elements, I have assumed that He is always neutral, N and O are tied to H through charge exchange and C, S, Si, Fe and Mg are always singly ionized.

In the initial configuration, at $t = 0$, one has:

$$\mathcal{H}_0 - \mathcal{L}_0 = 0.$$

In order to examine the effects of different assumptions for the heating term, calculations have been carried out setting: i) $\mathcal{H}(t) = \mathcal{H}_0$, ii) $\mathcal{H}(t)$ proportional to the particle density n, i.e. $\mathcal{H}(t) = n\,\mathcal{H}_0$, and iii) $\mathcal{H}(t) = 0$. One notices that the first two initial configurations considered are also equilibrium configurations, while in the third case the unperturbed medium cools steadily on a time scale t_{rad} (see below). The evolution time t is measured in units of $t_{cr} = a/c_s$ (a is the initial radius and c_s is the isothermal sound speed) and the velocities in units of c_s.

The parameters that control the dynamics of the system (see RBMF) are the (internal) jet Mach number M, the initial density ratio $\nu \equiv n_\infty/n_0$, with $n_\infty \equiv n(r = \infty)$ and $n_0 \equiv n(r = 0)$. The radiation effects instead are fully determined setting the initial on-axis temperature $T_0 \equiv T(r = 0)$ and cooling time $\tau_{\rm rad} \equiv t_{\rm rad}/t_{\rm cr}$, with $t_{\rm rad} = p/[(\Gamma - 1)\mathcal{L}]$, or equivalently the column density $n_0 a$. As far as the heating time scale $\tau_{\rm heat}$ is concerned, one can note that, being this parameter initially set equal to $\tau_{\rm rad}$, it is not an independent parameter. As discussed in RBMF, the initial parameters appropriate for the application to stellar jets yield $\tau_{\rm rad} \gg 1$, i.e. the initial conditions are nearly adiabatic. As unstable modes grow and shocks form, radiation effects become important locally in the compressed regions. In the calculation I have set $M = 20$, $\nu = 1$, $T_0 = 10,000$ K and $n_0 a$ such that $\tau_{\rm rad} = 131.2$.

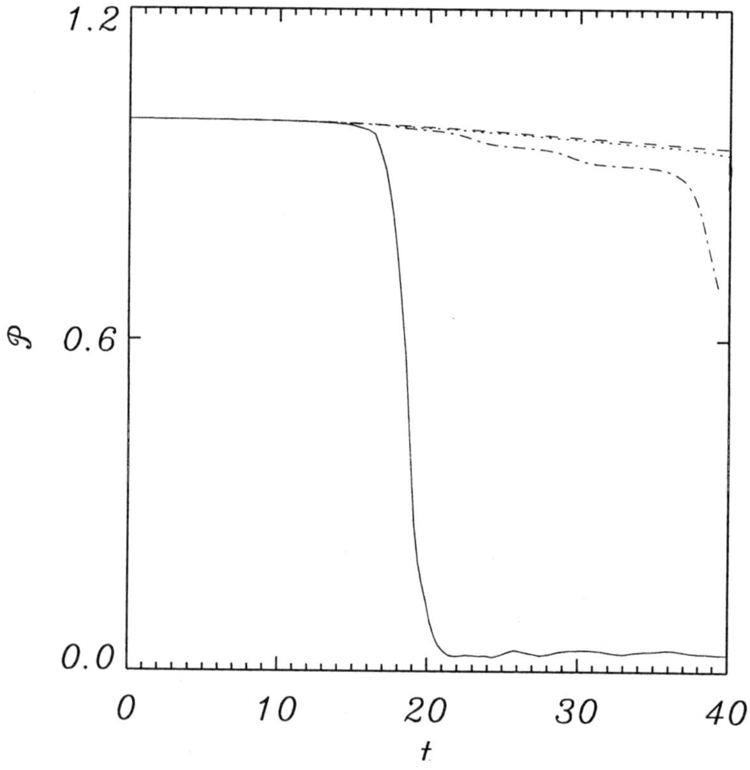

FIGURE 1. Jet's momentum as a function of time: adiabatic conditions (solid line), $\mathcal{H} = \mathcal{H}_0$ (dotted line), $\mathcal{H} = n\mathcal{H}_0$ (dashed line), and $\mathcal{H} = 0$ (dot-dashed line).

3. RESULTS

I focus the attention to the general dynamical properties of the instability evolution, comparing the momentum deposition and the entrainment properties under the different assumptions for the heating term. As in RBMF, the longitudinal momentum carried by the jet is defined:

$$\mathcal{P} = 2\pi \int_0^D \int_0^R \mathcal{T}\rho v_z r\,dr\,dz$$

where $\mathcal{T}$ is a tracer that distinguish between the jet matter from the external one. The decrease of $\mathcal{P}$ in time equals the momentum deposition in the ambient gas.

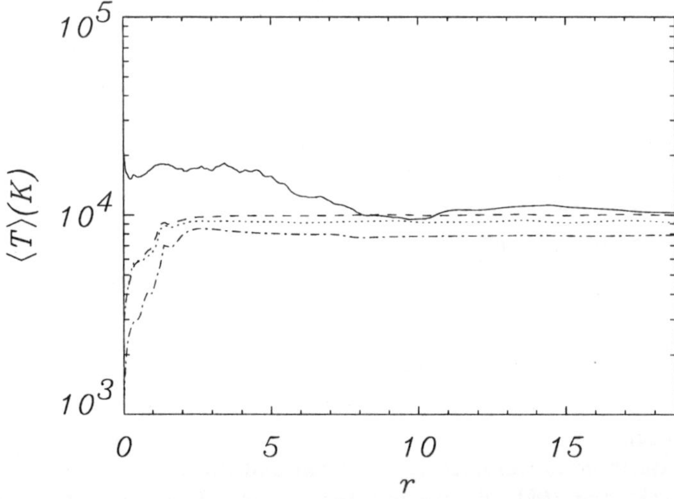

FIGURE 2. Longitudinal-mean of temperature as a function of r at $t = 30$: adiabatic limit (solid line), $\mathcal{H} = \mathcal{H}_0$ (dotted line), $\mathcal{H} = n\mathcal{H}_0$ (dashed line), and $\mathcal{H} = 0$ (dot-dashed line).

Figure 1 summarizes the general behavior of the jet's momentum as a function of time. One notices (as in RBMF) the dramatic effects of radiative losses on the instability comparing the results obtained in adiabatic conditions (solid line) with those including radiation: at $t \simeq 16$ the shocks, originated by the Kelvin-Helmholtz modes in the (adiabatic) jet, begin to produce a strong mixing that entrains external matter and deposit most of the jet's momentum into the ambient; on the contrary, radiative jets tend to maintain the largest fraction of their momentum. Examining Fig. 1 in more detail, one notes that the cases of constant heating (dotted line) and heating proportional to n (dashed line) are almost not distinguishable, while the jet loses a larger fraction of its initial momentum in the case of pure cooling (dot-dashed line).

This behavior can be understood looking at Fig. 2, where I plot the temperature, averaged on the longitudinal direction, as a function of the radial coordinate r at $t = 30$. It is transparent the qualitative difference of the adiabatic case (solid line) with respect to the radiative ones: the jet's region, overheated by strong adiabatic shocks, has expanded entraining external matter and losing the longitudinal momentum, while in case of radiative jets cooling prevails, in average, over shock heating preventing expansion and mixing. Looking at Fig. 2 in more detail, one notes that the unbalanced cooling of the external medium decrease its mean temperature (dot-dashed line), favoring a moderate expansion and mixing, in agreement with a more pronounced decrease of the jet's momentum seen in Fig. 1.

Figure 3 shows the (logarithmic) density distribution, at $t = 30$, for the cases examined. One can notice how the cases of heating constant and proportional to n behave in the same way, and that the case of cooling jet undergoes a moderate expansion, but much less than the adiabatic jet.

A picture of the entrainment process in the regions closer to the jet, where most of its momentum is deposited, is given in Fig. 4 that show, at different times, the distribution of the external material as a function of the longitudinal velocity. To construct these figures I have divided the velocity interval between $v_z = -2$ and $v_z = V_0$ (with $V_0 \equiv V(r = 0, t = 0)$ in 30 bins and plotted the histograms of the fractions of material found in each velocity bin and initially not belonging to the jet, integrated up to a radius $r_e = 4a$. The different panels in Fig. 4 refer to the adiabatic case and to all the cases examined for radiative jets. The distributions at the two different times are shown with different gray shades and the later time is superimposed on the earlier time. One notes that in the adiabatic case one has, at $t = 30$, a fraction of external matter, initially at rest, that has been accelerated up to $v_z = 4$ by entrainment processes; while Fig. 4 confirms that radiative jets do not entrain external medium noticeably.

4. SUMMARY AND CONCLUSIONS

I have considered the non-linear evolution of Kelvin-Helmholtz instabilities in radiative jets. I have discussed in particular the effects of different assumptions for the heating $\mathcal{H}(t)$, namely: i) $\mathcal{H}(t) = \mathcal{H}_0$,

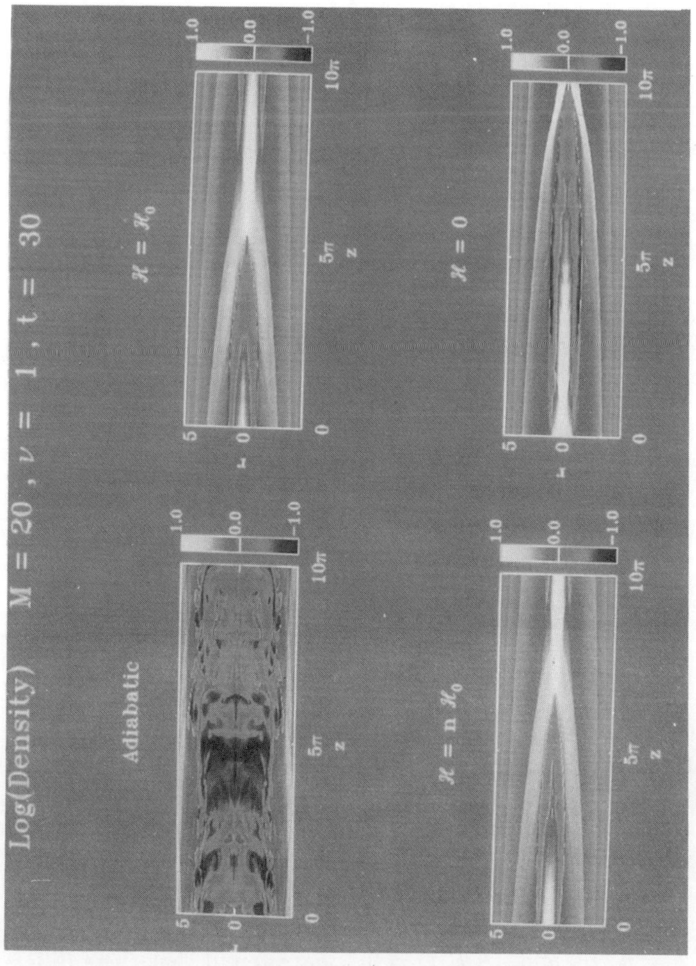

FIGURE 3. Color image of the (logarithmic) density distribution, at $t = 30$, for the cases considered (see Color Plate 5).

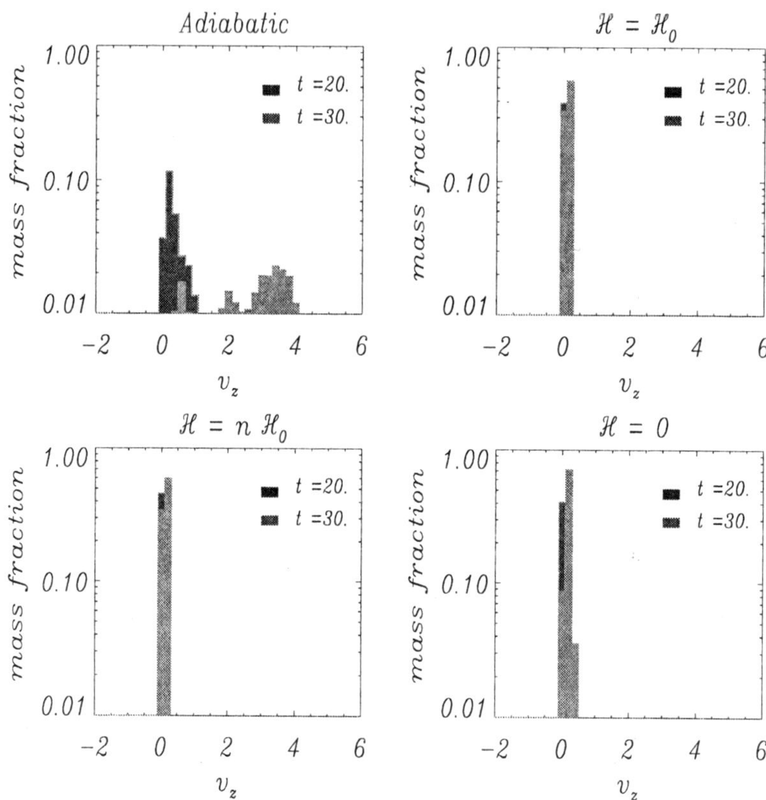

FIGURE 4. Histograms of the mass fraction of the material, initially not belonging to the jet, found at different velocities for the adiabatic and all cases examined. Note that the mass fraction scale is logarithmic in order to better represent the tail of the distributions.

ii) $\mathcal{H}(t) = n\,\mathcal{H}_0$, and iii) $\mathcal{H}(t) = 0$. I have found that these different assumptions do not modify the *qualitative* behavior of the instability,i.e. axially symmetric radiative jets can propagate for long distances into the molecular cloud environment without being disrupted by instabilities.

However, when axial symmetry is removed and the full 3-D analysis is carried out in adiabatic conditions (see Bodo Chapter 15), asymmetric modes play a very important role in the jet's mixing and matter entrainment. It remains to be seen whether the inclusion of radiation continues to play a stabilyzing role also in the case of, the more realistic, 3-D case.

REFERENCES

Bally, J., & Devine, D., 1994, *Ap.J. Lett.*, **428**, L65.

Blondin, J.M., Fryxell, B.A., & Königl, A., 1990, *Ap.J.*, **360**, 370.

Bodo, G., Massaglia, S., Ferrari, A., & Trussoni, E., 1994, *A&A*, **283**, 655.

Bodo, G., Massaglia, S., Rossi, P., Rosner, R., Malagoli, A., & Ferrari, A., 1995, *A&A*, 303, 281.

Colella, P., & Woodward, P.R., 1984, *J. Comp. Phys.*, **54**, 174.

Massaglia, S., Rossi, P., Bodo, G., & Ferrari, A., 1996, *Ap. Lett. Comm.*, **34**, 295 (MRBF).

Massaglia, S., Micono, M., Ferrari, A., Bodo, G., & Rossi, P., 1997, in *Herbig-Haro Flows and the Birth of Low Mass Stars*, eds. B. Reipurth & C. Bertout, Kluwer Academic Press, p.335.

Raga, A.C., Taylor, S.D., Cabrit, S., & Biro, S., 1995, *A&A*, **296**, 833.

Rossi P., Bodo G., Massaglia S., Ferrari A., 1997, *A&A*, **321**, 672 (RBMF).

Stone, J.M., 1997, in *Herbig-Haro Flows and the Birth of Low Mass Stars*, eds. B. Reipurth & C. Bertout, Kluwer Academic Press, p.323.

Stone, J.M., Xu, J., & Hardee, P.E., 1997, *Ap.J.*, **483**, 136.

CHAPTER 21

KNOT FORMATION IN HERBIG-HARO JETS

M. MICONO

Dipartimento di Fisica Generale dell'Università, Via P. Giuria 1, I-10125 Torino, Italy

We present the results of a numerical simulation of the spatial evolution of the Kelvin-Helmholtz instability in a hydrodynamical jet with non equilibrium radiative losses. Morphological features, that result from the instability evolution, such as knots shapes and spacing are described, lines flux ratios and knots proper motions are calculated, and a comparison of the expectations of the IWSs and the KH models on observable quantities is performed.

1. INTRODUCTION

One of the main issues on which theoretical work on stellar jets concentrated in the last years is trying to explain how emission knots, the main feature common to most Herbig-Haro jets, form. While it is generally acknowledged that the radiative emission from Herbig-Haro jets is shock wave radiation, it is not yet clear how shocks can arise along the jet; two models have been proposed: the internal working surfaces (IWSs) model, and the model involving Kelvin-Helmholtz (KH) instability.

According to the IWSs model, knots in jets form as a result of variations in the velocity of material ejection from the source. While travelling downstream, advected by the outflow, perturbations steepen into discontinuities giving rise to shock waves inside the jet. Analytical calculations have been performed by Raga and Kofman (1992) and Kofman and Raga (1992) who derived relations connecting the observed quantities such as shock velocities, Hα luminosities, knots separation and proper motions at an asymptotical distance from the source, to the amplitude and the period of variations in the velocity of gas ejection. More recently, Stone and Norman (1993) and Biro and Raga (1994) performed multidimensional numerical simulations where they studied the general behavior of a cooling jet with supersonic variations in the magnitude of the outflow velocity, and the structure and the temporal and spatial evolution of the internal working surfaces that form.

The first linear studies on the KH instability in an astrophysical jet, were motivated by the necessity of explaining how could a jet survive instability for distances as large as those observed in AGN jets; see for example the linear stability analysis by Ferrari et al. (1978), Hardee, (1979), Hardee and Norman (1988). Bührke et al. (1988) suggested that the nonlinear evolution of the axisymmetrical modes of the KH instability could be responsible of the formation of the observed knots in stellar jets. This prediction was confirmed by the numerical calculation of Bodo et al. (1994), who studied the long term evolution of a jet under the nonlinear development of the KH instability and by Rossi et al. (1997) who adopted the same temporal approach to study the effects led by the presence of a non-equilibrium cooling on the instability. The aim of this paper is to analyze the results of numerical simulations of a cooling jet under the action of the KH instability, addressing in particular to those morphological, kinematical and spectral features whose values and trends are known from observations. For this purpose we adopt a spatial approach, i.e. we study how a perturbation imposed near the source grows in a non linear fashion as it is advected downstream by the fluid, as a results of the rise of KH instability.

2. OBSERVATIONAL CONSTRAINTS

Knots in stellar jets are observed mainly in the hydrogen Balmer lines (especially Hα) and in the low-excitation forbidden lines [SII] ($\lambda\lambda$=6716,6731), [OI]($\lambda\lambda$=6300,6363), [NI]($\lambda\lambda$= 5198,5200). A very important quantity is the line ratio [SII]/Hα which is generally greater than unity, indicating that the excitation of the gas must be low, and thus that the shock velocity can't exceed the value of 50 km s^{-1}. For HH34 the value of the ratio ranges from 1 to 4, while in HH111 it is lower by a factor 1.5 - 2 (Hereafter, the data for HH34 and HH 111 are from Bührke et al. (1988), Reipurth (1989), Eislöffel and Mundt (1992), Heathcote and Reipurth (1992), Morse et al. (1993)) Lines observed are also a constraint on the maximum temperature that can be reached in shocks.

It can be noteworthy to mention the spatial trend of the peak emission, which in HH34 is observed to increase with the distance from the source, to reach a maximum in the middle of the chain and then to decrease. Regions where no emission is detected can be present in some sections of the entire length between the source and the final bow-shock; HH34 and HH111 show, for example an initial emission gap, and the first knot is detected at a distance of about 17 (HH34) and 36 (HH111) jet radii from the source. In HH30, instead, the emission begins from very near the central source (Lopez et al. (1995)).

The 10-13 knots detected in HH34 are very regularly spaced and the distance between two consecutive knots ranges from 2 to 7 jet radii, while in HH111 it ranges from 3 to 10 jet radii. In HH34 the farthest knots from the source, whose distance from the preceding one is grater compared to the medium spacing between knots in the chain, appears clearly bow-shaped; recent HST observations (Ray et al. (1996)) show that also knots displaced along the chain have such widened shape.

Another important measured feature is the proper motions of the knots; the ratio between knots velocity and jet's gas velocity is found to be between 0.4 and 1., and the highest values are attained by the farthest knots from the source.

3. RESULTS

The hydrodynamical equation of mass, momentum and energy conservation for a perfect gas were solved using a PPM code, and non equilibrium radiative losses were included in the computations by means of a time-splitting technique as described in Rossi et al. (1997). We worked in cylindrical geometry, with the jet axis placed on the bottom side of a 2500×248 grid, covering a domain of 664×64 jet radii; the boundary condition on the axis was a reflecting condition, while on the left side we had inflow conditions, with a perturbation imposed on the transversal velocity, and on the top and right sides of the grid the variables were forced to flow out with zero gradients. The control parameters are set as follows: for the Mach number we chose $M = 10$; the external to jet density ratio is $\nu = 1$, the initial temperature

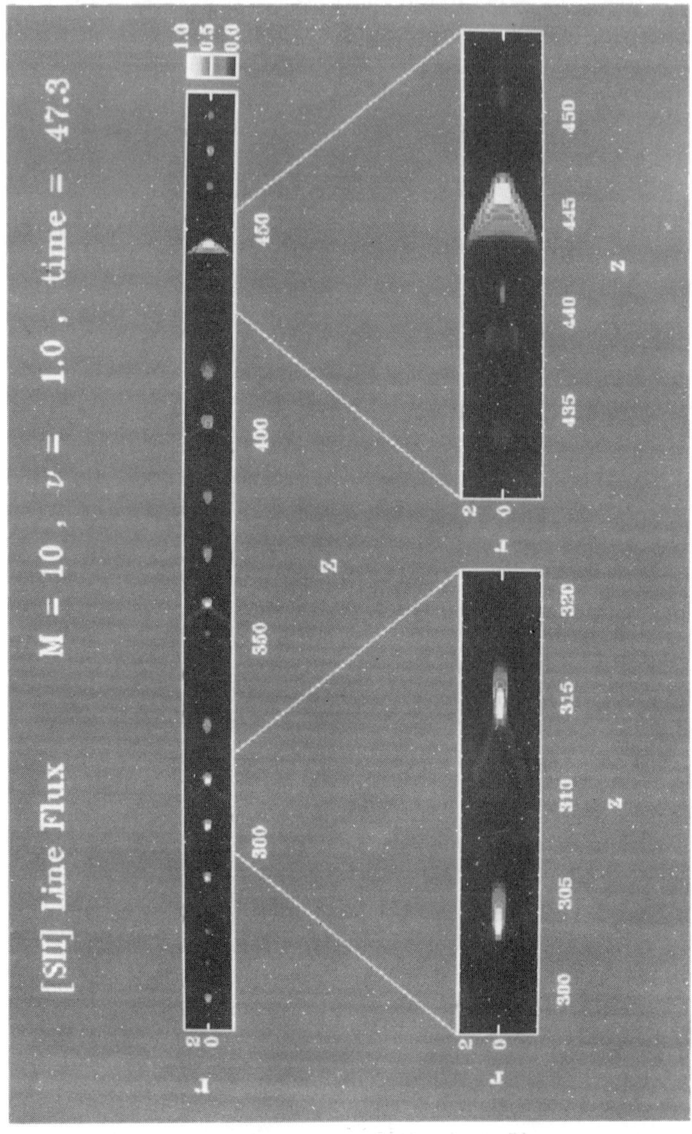

FIGURE 1. False-colors image of the [SII] lines flux at the time corresponding to 38.9 sound crossing times. The first section (linear phase) of the jet, with no emission, is not shown (see Color Plate 6).

on the jet axis is $T_0 = 10^4$ K, and the column density is fixed to $n_0 a = 10^{18}$ cm^{-2}. (For a detailed discussion see Micono et al. 1997).

In Fig. 1 a false-colors image of the [SII] lines flux at the time corresponding to 38.9 sound crossing times is presented. The observed structures are biconical shock waves centered on the jet axis; we can see that about 15 well defined knots have formed and that one of them, as a result of shock merging processes, shows a wide, bow-like shape. The first section of the jet, not shown in Fig. 1, is dominated by the linear phase of the instability evolution, no shocks and thus no emission are present.

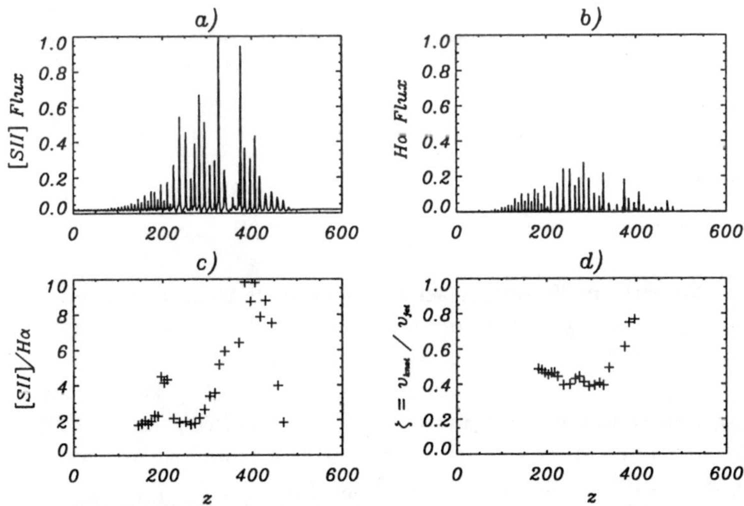

FIGURE 2. Profiles of [SII] and Hα fluxes (in arbitrary units) on the jet axis versus the distance from the source (panels a and b); [SII]/Hα line intensity ratio (panel c) and knots proper motions (panel d).

The separation between two consecutive knots is quite regular, and ranges from 3 to 6 jet diameters, and the spacing between the enlarged shock and the preceding one is greater than the average value.

In Fig. 2a,b the profiles of the [SII] and Hα fluxes on the jet axis are shown, from where we can deduce the trend of the radiation intensity along the jet, lower near the source and in the farthest regions. From Fig. 2c we can see how the values of the [SII]/Hα lines ratio are generally greater than unity, and how this value increases with distance from the source, reaches a maximum and then decreases.

The values and the trend of knots proper motions can be inferred from Fig. 2d: the ratio ζ between knot pattern speed and jet gas speed shows small variations around the value $\zeta \sim 0.4$ in the first section of the jet and increases up to $\zeta \sim 0.8$ for the farthest knots from the source.

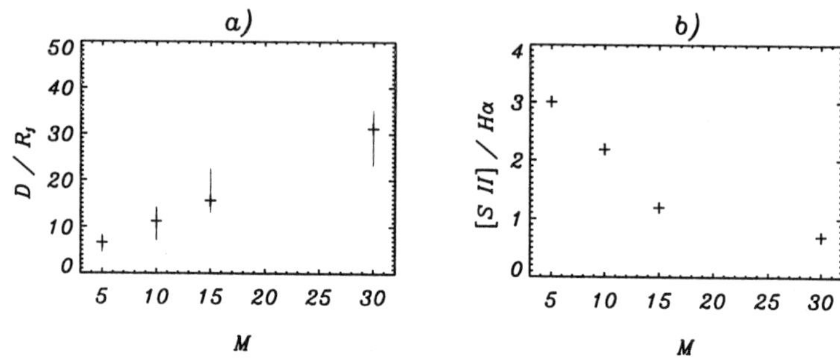

FIGURE 3. Plot of the mean spacing between knots vs Mach number (panel a) and plot of the [SII]/Hα lines ratio vs M.

We performed another set of calculations to study the effect of varying the Mach number of the jet on the observational quantities, and the results are shown in Fig. 3a and 3b. As the Mach number increases, the intra-knot spacing increases (Fig. 3a), as was already suggested from the linear analysis of Hardee and Norman (1988); the [SII]/Hα lines ratio is lower for higher Mach number, suggesting a higher excitation in the jet's gas (Fig. 3b).

4. SUMMARY AND CONCLUSIONS

To conclude, we recall the results of our model and compare them with the results obtained in previous analytical studies and numerical simulation of jets with a variable injection velocity, concentrating upon the features which can have observational constraints. As regards the knots morphology, the IWSs models leads to bow-shock like morphologies for every structure which forms; an interesting result of our calculation is that bow-shock like morphologies can appear also as a consequence of the development of the KH instability, for some knots, as a result of shock merging at advanced stages of the evolution.

The IWSs model doesn't lead to an emission gap between the source and the first knot, since the steepening of the profile in the velocity perturbations into shocks is very fast; indeed the line intensity profiles show an abrupt increase near the central source and then a smoother decline (Biro and Raga, 1994). A jet which undergoes the development of the KH instability, instead, cannot show emitting patterns in the section nearest to the source, where shocks have not yet formed; the intensity of the first shocks that form is low, then increases, reaches a maximum in the middle of the chain and decreases in the farthest knots.

In order to admit the low shock velocities required by the low excitation lines observed, the speed of the internal working surface cannot be less than 90% the jet gas speed. As we have seen, instead, the KH model leads to knots moving with a velocity which varies from 40% to 80% the jet gas velocity, with a well defined trend along the jet axis.

The Kelvin-Helmholtz instability is a general hydrodynamical phenomenon which cannot be avoided if a jet is in pressure equilibrium with the external medium, and a variation in the ejection velocity could be seen as an initial perturbation which can evolve in a non linear fashion as the instability grows. Characteristic features do not depend on the details of the velocity variations, such as the period and the amplitude of the pulses, but only on jet's parameters, i.e. density ratio and above all Mach number. In the KH model, besides, two important features, the line intensity ratio and the knot spacing are connected each other as both depend on the Mach number of the jet; an observational evidence of such a behavior could be inferred by a comparison of the two jets IIII34 and HH111.

REFERENCES

Biro, S., & Raga, A.C., 1994, *Ap.J.*, **434**, 221.

Bodo, G., Massaglia, S., Ferrari, A., & Trussoni, E., 1994, *A&A*, **283**, 655.

Bührke, T., Mundt, R., & Ray, T.P., 1988, *A&A*, **200**, 99.

Eislöffel, J., & Mundt, R., 1992, *A&A*, **263**, 292.

Ferrari, A., Trussoni, E., & Zaninetti, L., 1978, *A&A*, **64**, 43.

Hardee, P.E., 1979, *Ap.J.*, **234**, 47.

Hardee, P.E., & Norman, M.L., 1988, *Ap.J.*, **334**, 70.

Heathcote, S.R., & Reipurth, B., 1992, *A.J.*, **104**, 2193.

Kofman, L., & Raga, A.C., 1992, *Ap.J.*, **390**, 359.

Lopez, R., Raga, A.C., Riera, A., Anglada, G., & Estalella, R., 1995, *MNRAS*, **274**, 19L.

Micono, M., Massaglia, S., Bodo, G., Rossi, P., & Ferrari, A., 1997, *A&A*, in press.

Raga, A.C., & Kofman, L., 1992, *Ap.J.*, **386**, 222.
Ray, T.P., Mundt, R., Dyson, J.E., Falle,S.A.E.G., & Raga, A.C., 1996, *Ap.J.*, **468**, L103.
Rossi P., Bodo G., Massaglia S., Ferrari A., 1997, *A&A*, **321**, 672.
Stone, J., & Norman, M., 1993, *Ap.J.*, **413**, 210.

CHAPTER 22

A NEW MODEL FOR GAS EXCITATION IN THE BEAM OF STELLAR JETS

F. BACCIOTTI[1], A. POUQUET[2],
C. CHIUDERI[1]

[1]*Dipartimento di Astronomia e Scienza dello Spazio, Univer-
sitá di Firenze, Largo E. Fermi 5, I-50125 Firenze, Italy*

[2] *Observatoire de la Côte d'Azur, CNRS URA 1362, BP 4229,
06304 Nice–Cedex 4, France.*

We discuss a new mechanism for the excitation of the optically emitting gas
in the beam of protostellar jets, based on the stability properties of Alfvén
waves produced by the rotation of the accretion disc and propagating along
the jet beam. We adopt a magnetohydrodynamic (MHD) description including the Hall term, as appropriate for the partially ionized gas filling
the jet channel, and assume quasi-equipartition of thermal and magnetic
energies. When instability develops shock-like features appear in the flow
which generate heating and density enhancements in both lobes of the jet,
depending on the alignment between the angular momentum vector of the
disc and the direction of the background magnetic field.

1. INTRODUCTION

The mechanism of formation of the regular chain of weak traveling shocks required to explain the emission observed in the beam of

Herbig-Haro (HH) jets is still under debate (see, e.g., Raga and Kofman 1992, Stone and Norman 1993, Bodo et al. 1994). An attractive possibility relies on the inclusion of magnetic effects, that may influence the formation and properties of shocks occurring in the flow. In most star formation scenarios the young star and its accretion disc are embedded in a hourglass-shaped magnetic field. Magneto-centrifugal effects have been recently invoked as the dominant mechanism for the acceleration and the collimation of bipolar outflows (see, e.g., the review by Königl 1996). Here we illustrate a recent model (Bacciotti, Chiuderi and Pouquet 1997, hereafter BCP) in which the interplay between disc rotation and magnetic effects is also responsible for the gas excitation and the nodular morphology of the jet.

We consider a situation in which the magnetic field in the jet is strongly tied to the accretion disc and is continuously wound by the disc rotation. The magnetic twist propagates along the large scale field parallel to the axis of the jet in the form of Alfvén waves of constant amplitude transported by the fluid motion. When the magnetic stresses that accumulate by effect of the rotation overcome a critical threshold they are released producing an amplitude modulation of the waves, which may steepen as the wave propagates along the jet. In this circumstance energy may be transferred to the fluid, and shock fronts that locally heat the gas may appear. If the process of build up and release of magnetic stresses occurs in a periodic fashion, a chain of regularly spaced shocks will travel in the flow. In this preliminary study we concentrate on the propagation aspects, in order to verify if shock-like features may actually form in the flow.

2. MODEL EQUATIONS

Recent diagnostic investigations (Bacciotti et al. 1995, 1996, Bacciotti and Eislöffel, 1997) proved that the gas in the beam of HH jets is partially ionized ($\xi = n_{H+}/n_H \approx n_e/n_H \sim 0.1 - 0.3$, where n_{H+} and n_H are the number densities of ionized and neutral hydrogen, respectively). As a consequence the so-called Hall term must be included in the MHD equations due to the fact that the current density is enhanced by the different slowing down of the two charged species in their collisions with neutrals (see BCP for details). The Hall term, whose importance increases inversely with the ionization fraction, introduces dispersion which competes with nonlinearity, thus reducing the strength of shock fronts that may form in the medium. As an energy equation to close the pertinent Hall-MHD system we adopted a polytropic law $d(P\rho^{-\lambda})/dt = 0$, with a variable parameter λ, that is set either to 5/3 in the adiabatic case, or to 1/3 when mimicking heating and radiation (Passot et al. 1995). We study the propagation of waves in close proximity of the jet axis, thus allowing for a dependence on the axial coordinate x only, although keeping the three components of vectorial fields. We assume strong coupling between magnetic and fluid effects, which traduces in the condition $\beta = P_{\text{gas}}/P_{\text{magn}}$ close to unity; this implies that the average intensity of the background field (not directly measurable in HH jets; see however Ray, Chapter 16) is about 200 μG, which, in turn, gives an Alfvén speed c_A of about 5 km s^{-1} (BCP). The regime in which the defocusing effects associated with the dispersive Hall term compete with the nonlinear evolution is conveniently studied expanding both variables and coordinates in terms of a small parameter ϵ, that is a measure of the amplitude of the carrying wave normalized to the background field B_0 ('reductive perturbation expansion', e.g. Hada 1993):

$$\rho = 1 + \epsilon\rho_1 + \epsilon^2\rho_2 + \dots \quad ; \quad P = 1 + \epsilon p_1 + \epsilon^2 p_2 + \dots$$

$$u = u_0 + \epsilon u_1 + \epsilon^2 u_2 + \dots \quad ; \quad v \equiv v_y + iv_z = \epsilon v_1 + \epsilon^2 v_2 + \dots.$$

$$h = 1 + \epsilon h_1 + \epsilon^2 h_2 + \dots \quad ; \quad b \equiv b_y + ib_z = \epsilon b_1 + \epsilon^2 b_2 + \dots$$

$$X = \epsilon(x - Ct) \quad ; \quad \tau = \frac{1}{2}\epsilon^2 t \quad ; \quad \beta \approx 1 \;\Rightarrow\; c_s^2/c_A^2 - 1 = \epsilon\Delta.$$

In the above expressions ρ is the total density (ions + neutrals) normalized to the average jet density ρ_0, $\mathbf{B} = (h, b_y, b_z)$ is the magnetic field normalized to the background field B_0, P is the total pressure normalized to $B_0^2/4\pi$, $\mathbf{U} = (u, v_y, v_z)$ is the velocity of the center of mass of the fluid particle normalized to c_A, c_s is the average sound speed, Δ is a parameter of order unity, and C is the constant speed of the moving frame in which it is convenient to work. Time and space are normalized to Ω_i^{-1} and c_A/Ω_i, where $\Omega_i = eB_0/m_i c$ is the ion gyro-frequency. Compatibility conditions yield $u_0 - C = \sigma = \pm 1$, where $+(-)$ stands for regressive (progressive) propagation with respect to the orientation of B_0; the reference frame practically moves with the bulk flow. The above expansion reduces the full Hall-MHD system to the following equations for the first-order quantities:

$$\sigma\frac{\partial\rho_1}{\partial\tau} = \frac{\partial}{\partial X}\left(\left(\frac{\lambda+1}{2}\right)\rho_1^2 + \Delta\rho_1 + \frac{|b_1|^2}{2}\right) \tag{1}$$

$$\frac{\partial b_1}{\partial\tau} = \sigma\frac{\partial}{\partial X}(\rho_1 b_1) - \frac{i}{\xi}\frac{\partial^2 b_1}{\partial X^2} \tag{2}$$

$$u_1 = -\sigma\rho_1, \quad \rho_1 = p_1, \quad b_1 = \sigma v_1. \tag{3}$$

These equations describe the behavior of small amplitude magnetic and density perturbations in a moderately ionized medium with constant ionization fraction, under the combined influence of nonlinearity and dispersion.

3. THE MODULATIONAL INSTABILITY

Circularly polarized Alfvén waves of constant amplitude propagating in a uniform medium are exact solutions provided that $\omega_0 = -k_0^2/\xi$, where k_0 and ω_0 are the wavenumber and the frequency of the carrying wave, respectively. However, any small deviation from a perfectly constant amplitude induces a dynamical reaction in the system due to the coupling between fluid and magnetic effects. A linear stability analysis of Eqs. (1) -(3) in the limit of long wavelength perturbations shows that at fixed Δ the wave becomes unstable only for one sign of k_0. The sign of k_0 (i.e. the polarization character of the wave) is determined by the sense of rotation impressed by the disc to the transverse component of the magnetic field. The perturbed wave envelope will be unstable only when the rotation of the transverse component agrees with the sense of the thermal ion motion around the background magnetic field, since in that case a resonant amplification sets in. Due to the hourglass configuration of the large-scale magnetic field, the waves traveling away from the star/disc system in the two opposite jet lobes show analogous stability/instability properties if Δ has the same sign on both sides, developing instabilities or not depending on the alignment between the angular momentum vector of the disc and the background magnetic field.

In order to follow the dynamical evolution of the system beyond the linear regime, we performed a numerical analysis of Eqs. (1)-(3) (see BCP for details). In Fig. 1 we show the results obtained for one stable and two unstable cases, corresponding to adiabatic ($\lambda = 5/3$) and 'radiative' ($\lambda = 1/3$) conditions, starting always from $\rho_1 = 0.0$ and a circularly polarized wave whose amplitude is subject to a 5% long-wavelength gaussian modulation. The ionization fraction was kept constant at $\xi = 0.1$. The development of the instability results in the formation of sharp fronts in the profile of the first order (in ϵ) density (and velocity) perturbation. As expected, the radiative case develops stronger fronts. The perturbation of the magnetic amplitude is also subject to growth, both in the adiabatic and in the radiative cases. In the stable case, on the contrary, the perturbations never grow very far beyond the initial imposed value.

The density maximum in the simulations corresponds to a bright knot, and the density minimum to the weakly emitting 'inter-knot' region. To attempt a first comparison with observations we calculated the average luminosity contrast between these regions predicted by our simulations, as a function of the 'smallness parameter' ϵ.

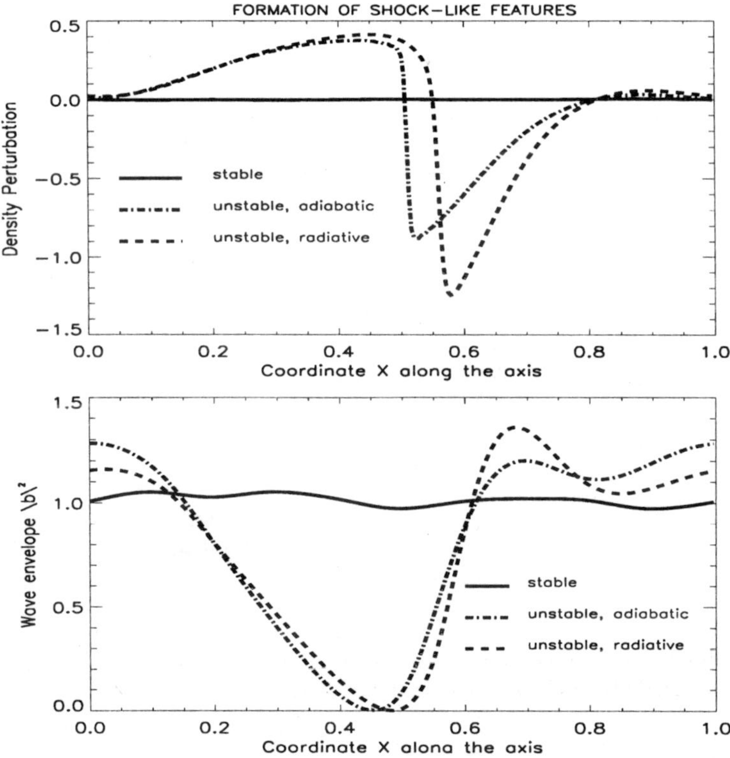

FIGURE 1. Top: density perturbation along the jet axis for two unstable cases ($k_0/\xi = -2; \Delta = 4$), one adiabatic and the other radiative, superimposed to the profile for the stable case ($k_0/\xi = 2; \Delta = 4$). Bottom: wave envelope $|b|^2 = b_{y1}^2 + b_{z1}^2$ for the same three cases.

The observed average value of the luminosity contrast (Mundt, Ray and Raga 1991) is reproduced with $\epsilon \sim 0.15$ in the adiabatic case, and $\epsilon \sim 0.23$ in the radiative one, i.e., when the amplitude of the carrying Alfvén wave continuously excited by the disc rotation is roughly 20% of the amplitude of the background field (see Fig. 2).

Our simulations show that in the unstable parameter regime sharp compression fronts do form in the flow which could account for the heating and the excitation of the gas. These results suggest that the nonlinear steepening of modulational perturbations of Alfvén waves could constitute a promising alternative to the pure fluid mechanisms so far proposed to explain the observed peculiarities of the optical emission in the beam of HH jets.

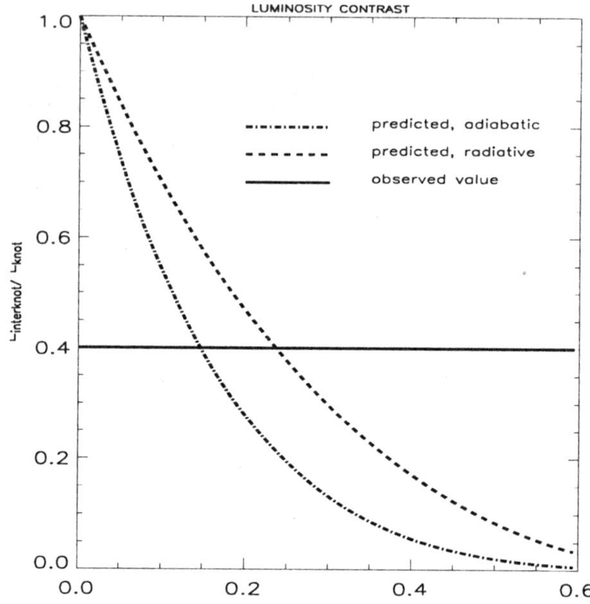

FIGURE 2. Luminosity contrast as a function of the smallness parameter ϵ for the adiabatic and radiative unstable cases.

REFERENCES

Bacciotti, F., Chiuderi C., & Oliva, E. 1995, A&A, **296**, 185.

Bacciotti, F., Hirth, G. A., & Natta, A. 1996, A&A, **310**, 309.

Bacciotti, F., Chiuderi, C., & Pouquet, A., 1997, ApJ, **478**, 594 (BCP).

Bacciotti, F. & Eislöffel, J., 1997, A&A, submitted.

Bodo, G., Massaglia, S., Ferrari, A., & Trussoni, E., 1994, A&A, **283**, 655.

Hada, T., 1993, Geophys. Res. Lett., **20**, 2415.

Königl, A., 1996, in Proc. of the M-P.I.A. Conference "Disk and outflows around young stars", Lecture Notes in Physics Vol. 465, S. Beckwith, J. Staude, A. Quetz & A. Natta eds., Springer Verlag Berlin Heidelberg, Germany, p. 282.

Mundt, R., Ray, T. P. & Raga, A. C., 1991, A&A, **252**, 740.

Passot, T., Vazquez, E., & Pouquet, A., 1995, *ApJ*, **455**, 536.
Raga, A.C., & Kofman, L., 1992, *ApJ*, **386**, 222.
Stone, J.M., & Norman, M.L., 1993, *ApJ*, **413**, 210.

CHAPTER 23

FINAL REMARKS

A. FERRARI[1] AND G. BENFORD[2]

[1]*Università and Osservatorio Astronomico, Torino, Italy*

[2]*Department of Physics and Astronomy, University of California, Irvine, USA*

Our present understanding of jets is based on the paradigm of accretion discs around a central more or less compact object, namely a black hole, neutron star, white dwarf or protostar, with matter falling onto the central object along the equatorial plane and expelled at supersonic, in most cases at relativistic, speed along the rotation axis. Although we still lack a complete model for connecting the various physical processes, the observational evidences in this direction appear very strong. In particular the images of HST of the discs and jets in the active galaxy M 87 and in HH 30 and HH 34 (Tom Ray, Chapter 16) adapt convincingly to the scheme proposed 30 years ago by Donald Lynden-Bell. Issues concerning confinement of the sources have been discussed by Edoardo Trussoni (Chapter 10). The relativistic galactic sources have been also analyzed by Wolfgang Brinkmann (SS 433, XRB, Chapter 19).

These observational evidences are summarized in Table 1 for the various classes of objects. We also include in our list the so-called cataclismic variables (CV) that may be interpreted in the same framework, as a disc inflow is clearly detected in their outbursts. In Table 2 we give instead a list of the best conjectured physical parameters according to the accretion disc and jet scheme.

TABLE 1: Observational evidences

OBJECT	BIPOLAR OUTFLOW	DISC INFLOW
AGN	yes	?
YSO	yes	yes
XRB	yes	yes
BHXRT	yes	?
SSS	yes	?
CV	?	yes

TABLE 2: Physical parameters of jets

	AGN	YSO	BHXRT	XRB	SSS	CV
β_j	1	10^{-3}	0.9	0.3	10^{-2}	10^{-2}
L_j (erg s^{-1})	10^{47}	10^{35}	10^{38}	10^{40}	10^{38}	10^{33}
$M_{core}/M_\odot$	$\geq 10^8$	1	1	$1-10$	1	1
R_{core} (cm)	10^{14}	10^{11}	10^6	10^6	10^9	10^9
β_{esc}	1	10^{-3}	0.9	0.3	10^{-2}	10^{-2}
R_{coll} (cm)	10^{17}	10^{16}	10^{11}	10^{11}	10^{11}	10^{11}
B (G)	10^4	1		10^9		10^6
L_j/L_{Edd}	≤ 1	≤ 1		≤ 1	≥ 1	≤ 1
Rel. core/jet	yes/yes	no/no	yes/?	yes/no	no/no	no/no

A number of the old and new successes of this model in interpreting the various objects has been reported in the meeting (see the contributions by Camenzind, Rossi, Micono, Martì, Gomez). However several intriguing issues are still unsolved notwithstanding the many important contributions. We resume here some of the critical points that have been mentioned during the meeting and we agree that still require our attentive consideration in the future.

1. *The energy problem.* As pointed out by Geoffrey Burbidge back in 1956 with reference to the extended extragalactic radio sources, the first class of objects providing evidence for large scale jets, the power in active galaxies and in jets is very large; if one takes into account the efficiency factors that intervene in defining the tiny amount of energy that is going into radiation out of the global dynamics of these objects, this last quantity is estimated to be at the very limit of what can be extracted from a maximally rotating Kerr black hole. In order to push to the highest values the radiation efficiency coherent processes have been proposed

and experimental evidence in their favor has been presented by Greg Benford (Chapter 8).

2. *The jet acceleration problem.* The original ideas about supersonic jet production from the intense radiation field in the central powerhouse have basically failed to explain relativistic jets, although may apply to star formation regions. General relativistic models for relativistic jets including electrodynamic effects have been so far only marginally explored in the original papers by Blandford, Lovelace, Znajek and Phinney. Some useful results have been obtained in connection with pulsars by Goldreich, Ruderman, Arons. Whether relativistic jets emerge from the accretion funnel as collimated, neutralized ionized beams or instead carrying currents is not completely understood yet, although the importance of magnetic fields has been convincingly discussed by Max Camenzind (Chapter 1). The situation is certainly less critical in the case of protostellar jets; the general consensus, confirmed by models and numerical simulations, is that acceleration is similar to that blowing the solar wind, hydro or MHD (Dick Henriksen, Cristophe Sauty, Kanaris Tsinganos).

3. *The physical components of jets.* Observations clearly show the existence of relativistic particles and magnetic fields in relativistic jets and of thermal plasma in jets from star forming regions. Most likely all these ingredients belong to all types of jets. In addition in relativistic jets the formation of electron-positron pairs has been proposed as due to energetic electrodynamic processes taking place in the strong magnetic fields of the central power house. An important issue would be to investigate whether jets are charge neutralized and/or carry currents to the very large structures.

4. *The instabilities of jets.* Collimated jets appear to be subject to all kind of fluid instabilities. Numerical 3D hydro simulations, discussed at the meeting by Gianluigi Bodo (Chapter 15) show that strong mixing and entrainment of external medium slow down the dynamics to the subsonic regime and completely blur the propagation features. The problem may be somewhat alleviated by the presence of magnetic fields, but really convincing MHD simulations covering time or length scales relevant to the astrophysical situations have not been performed. Another proposed approach to solve the question was the self-consistent creation of overpressured cocoons around jets to avoid their free expansion; however also in this case mixing and entrainment continue to be present. Most likely current carrying jets with large scale induced azimuthal fields would instead solve the problem, but, as we said before, the support of a current over large scales

outside the central source has to be ascertained.

5. *Particle reacceleration in relativistic jets.* Consistent calculations of radiation from relativistic/nonthermal jets are being undertaken, including also the effect of particle reacceleration via shock waves or turbulent modes. Several models were proposed in the past; now one tries to connect those processes with the dynamical evolution of the jet fluid that includes ions, electrons and electron/positron pairs. The problem is relatively straightforward for what concerns the spectral shape of emission; more difficult is to reach a result for the total number of accelerated particles.

6. *High-energy radiation from AGN and intraday variability.* The data presented here by Gabriele Ghisellini (Chaper 6) underline the still emerging importance of variability in understanding jets at the smallest scale, where emission arises. Rapid changes due to extreme relativistic effects and perhaps also due to new emission mechanisms can pose quite severe constraints on jet behavior. Mentioned in several discussions was recent data on a stellar jet (Rodriguez and Mirabel) which displays very high brightness ($\sim 10^{18}$ K) and moderate relativistic factors $\gamma \sim 2$, yet cannot be pointing nearly at us. This case may presage a way to study collective jet emission processes at close range, *i.e.* in our own galaxy.

We would like finally express our feelings after these three days of discussion. Jets have endured our study two decades now, yielding some, but not all of their secrets. Many of the outstanding problems are mysterious. We suffer from the *weather problem*; jets are hard to explain in their endless variety, so the best we can hope for is an understanding of their *climate*: the general features of their origin and environment.

In this regard numerical simulations, once hailed as the guiding light, have to our minds failed to shed illuminating glow over central issues. As several speakers noted, especially Gianluigi Bodo (Chapter 15) and Silvano Massaglia (Chaper 20), jets seldom survive more than a few acoustic crossing times. The 3D simulations suffer faster growing instability than the 2D cases; greater dimensionality is no help in understanding how jets manage to maintain direction and coherent morphology while expanding by orders of magnitude. (And remember, the simulated environment is smooth and benign, unlike the likely real jet surroundings.)

This suggests that perhaps simulations omit crucial physics. A notable lack is any dynamically ordering magnetic fields. Max Camenzind (Chapter 1) has shown over the last several years, and in further detail here, that magnetically confined jets survive instability better than jets confined solely by external pressure. Their work includes the effect of flexing of fields as instability develops, which siphons energy from growth. A further effect not included so far is the loading of a jet with a magnetically linked cocoon (which in turn carries the return current at large radii). This can stabilize by adding an inertial term, forcing the jet to drag added mass.

Fat cocoons are energetically favored because they require less energy to supply the return current, by increasing the number of electrons which flow. The product nv^2A is then minimized for constant nvA.

Another effect neglected so far is the stabilizing effect of "jitter" in the beam equilibrium. No real jet will have fixed current and radius; the central source will experience *weather* and the flow must adjust itself. Perhaps surprisingly, this helps stability; the continuous adaptation proceeds by launching Alfvèn waves, which are in the frequency range affecting growth, and perform a dissipative role. Such effects are important in classical magnetic pinches, as shown by both theory and experiment. Energy then flows from large scale morphological deformations into smaller wiggles in magnetic fields, and ultimately probably toe heat through reconnection. Such effects could be critical for jet stabilization, especially in the difficult transition from the central engine into the extra- galactic arena, where the *weather* will be worst.

Considering electrodynamics underlines the central failure of present MHD codes. They treat hydrodynamics as primary, so shocks and mixing are well represented; but issues of magnetic structure and ordering of the flow elude such codes. Perhaps a simple code including only electrodynamics could illuminate this area. The larger problem of integrating the full Maxwell equations with present MHD approaches would then await understanding of purely electrodynamic effects.

Clearly imaginative, innovative approaches must emerge if we are to fathom the many unsolved jet mysteries. Perhaps the theorists should hasten to catch up with the torrent of data now available.

COLOR PLATE 1. Pressure distribution at eleven epochs (0 to 200 R_b/c in steps of 20) after the introduction of a square-wave perturbation to the flow Lorentz factor for the jet model discussed in the text. Owing to the decreasing pressure in the atmosphere, the jet expands with a mean opening half-angle of ~ 0.4°. The simulation has been performed over a grid of 1600 x 80 cells, with a spatial resolution of 8 cells/R_b in both radial and axial directions. Jet boundary is shown with a black curve (see page 97).

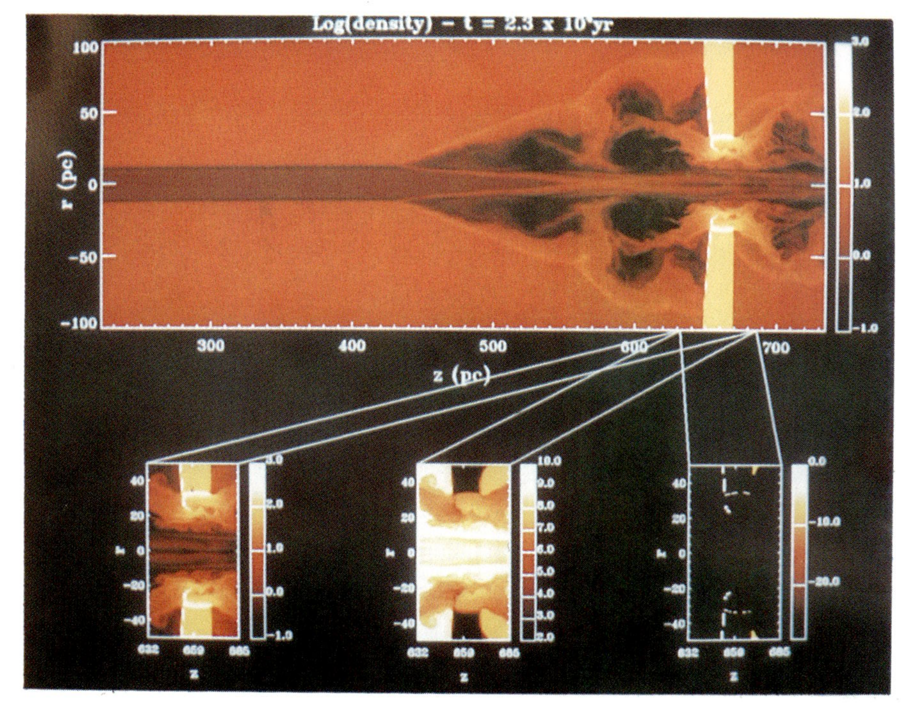

COLOR PLATE 2. The top panel shows a snapshot of *Log (n)* relative to jet/cloud interaction, while in the three low panels are shown, from left to right, the zoom of *Log(n)*, the *Log (T)* distribution and in the last one is a sort of *emission tracer* proportional to *Log (n^2)* for a temperature range between 5 x 10^3 and 2 x 10^4 K (see page 143).

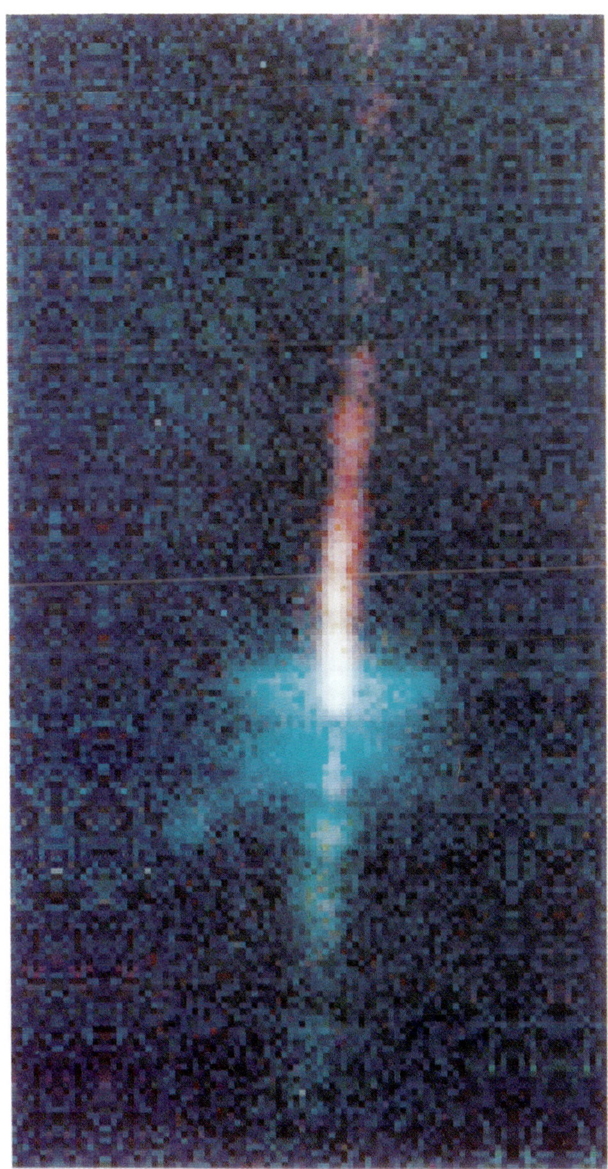

COLOR PLATE 3. HST image of HH 30 based on data from Ray et al. (1996). The jet has been oriented so that the blue-shifted flow points almost upwards. Blue light represents continuum emission, which in this case is scattered starlight from the 'top' and 'bottom' of the disk. The source itself is highly embedded and the disk can be seen in silhouette as the dark lane bisecting the nebular cusps. Red light is emission from the red [SII] doublet and green represents Hα. Note the asymmetry in excitation conditions in the flow and the counterflow. Image reconstruction done by C.R. O'Dell and S.V.W. Beckwith (see page 184).

COLOR PLATE 4. The three dimensional surface of the SS433 jet containing more than 60% of the original jet material after six precession periods (see page 214).

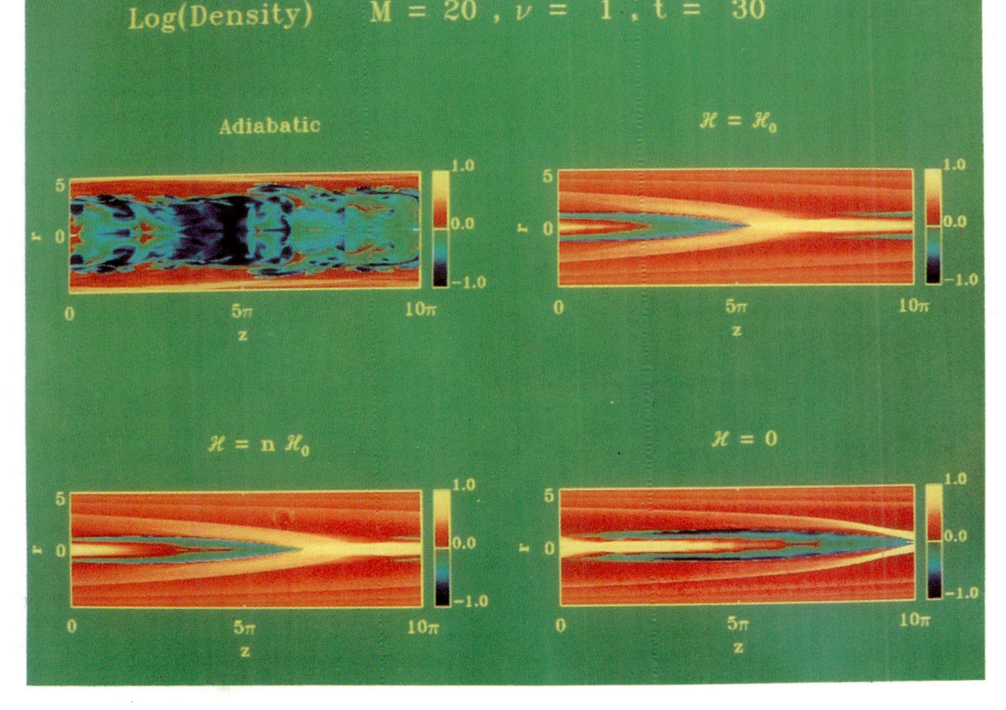

COLOR PLATE 5. Color image of the (logarithmic) density distribution, at $t = 30$, for the cases considered (see page 227).

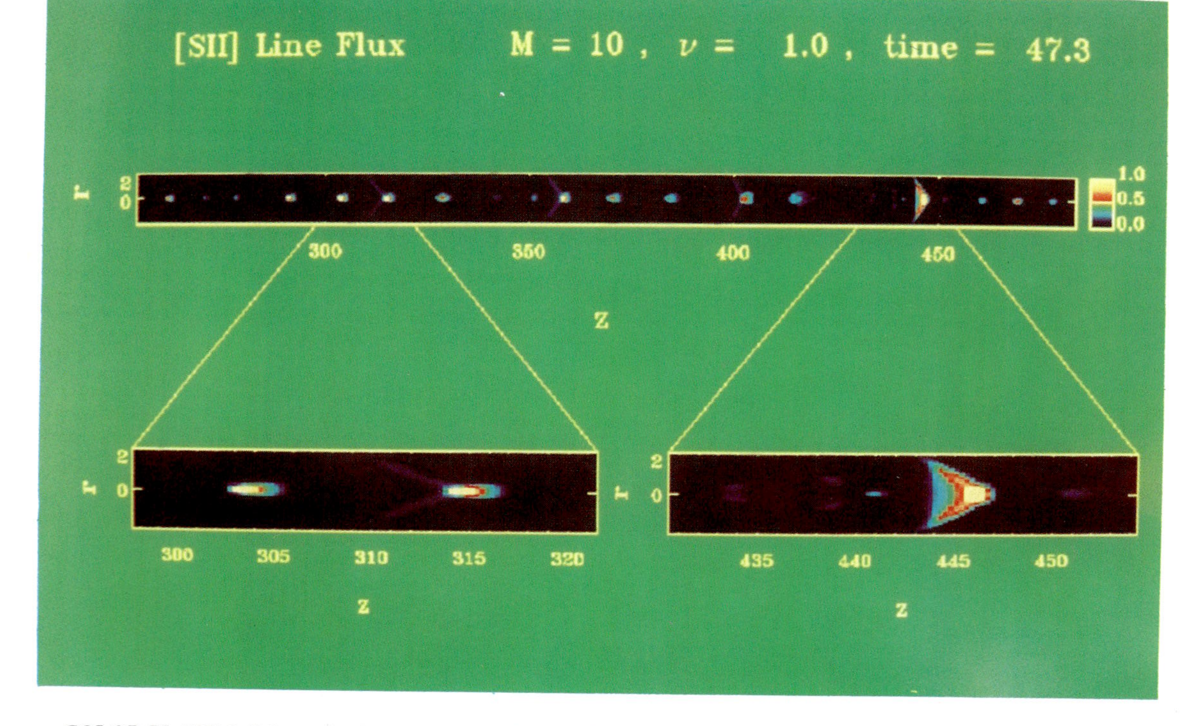

COLOR PLATE 6. False-color image of the [SII] lines flux at the time corresponding to 38.9 sound crossing times. The first section (linear phase) of the jet, with no emission, is not shown (see page 234).